The Oster Conspiracy of 1938

Also by Terry Parssinen

SECRET PASSIONS, SECRET REMEDIES:
NARCOTIC DRUGS IN BRITISH SOCIETY, 1820–1930

WEBS OF SMOKE: SMUGGLERS, WARLORDS, SPIES, AND THE
HISTORY OF THE INTERNATIONAL DRUG TRADE (with Kathryn Meyer)

The Oster Conspiracy of 1938

The Unknown Story of the Military Plot
to Kill Hitler and Avert World War II

TERRY PARSSINEN

HarperCollins*Publishers*

Grateful acknowledgment is made to the following for permission to reprint copyrighted material:

Dorothy Deutsch Thews, for permission to quote from Dr. Harold C. Deutsch Papers, Archives of the U.S. Army Military History Institute, Carlisle, Pa.

Topography of Terror Foundation, for permission to reprint the map "Das Regierungsviertel," from *Topography of Terror: Gestapo, SS, and Reichssicherheitshauptamt on the "Prinz-Albrecht-Terrain": A Documentation,* edited by Reinhard Rürup (Berlin, 1987), page 15.

Harold Ober Associates, for permission to quote from *The Diaries of Sir Alexander Cadogan* (New York, 1972), pp. 103, 105, 106–9.

Doubleday, a division of Random House, for permission to quote from Telford Taylor, *Munich: The Price of Peace,* copyright © 1979.

The University of York (England) for permission to quote from the Halifax Papers.

Nigel Nicolson for permission to quote from the Harold Nicolson manuscript diaries in Balliol Libary, Oxford.

FIRST EDITION

DESIGNED BY SARAH MAYA GUBKIN

Maps by Paul J. Pugliese and Lieutenant Colonel Winfried Heinemann, Ph.D.

Printed on acid-free paper

Library of Congress Cataloging-in-Publication Data

Parssinen, Terry M.
    The Oster conspiracy of 1938 : the unknown story of the military plot to kill Hitler and avert World War II / Terry Parssinen.—1st ed.
        p.   cm.
    Includes index.
    ISBN 0-06-019587-8
    1. Oster, Hans.   2. Hitler, Adolf, 1889–1945—Assassination attempts
3. Germany. Heer—Political activity.   4. Generals—Germany—Biography.
5. Germany—Politics and government—1933–1945.   I. Title.

DD247.O85 P37   2003
    943.086'092—dc21   2002068896

03   04   05   06   07   WB/RRD   10   9   8   7   6   5   4   3   2   1

For my father

Lieutenant Colonel Ollie Parssinen (U.S. Army, Ret.)

A member of the Greatest Generation

who fought for our freedoms

Hans Oster was a man after God's own heart,
of irreproachable character, great lucidity,
and iron nerve in the face of danger.

—FABIAN VON SCHLABRENDORFF,
Historian

This [German resistance] movement was not only not encouraged,
but was indirectly opposed by Germany's enemies. For anyone who
could not convince himself to become an emigrant, there remained
no other option, if one wished to live and die as a German patriot,
than to join this tenacious opposition. Fate and fault, fate and guilt,
are closely intertwined here on both sides. Only a time less torn by
the passions of the day will be able to judge whether the resistance
was truly useless or whether it laid the foundation from within
for the renewal of the German people.

—THEO KORDT,
Conspirator

Contents

Illustrations follow page 138.

Maps

Preface

The origin of this book goes back to a series of questions asked several years ago at the end of one of my classes on the History of the Second World War at the University of Tampa.

"Okay, that's it. Any questions? Stephanie?"

"Professor, when was the last chance that the Second World War could have been stopped?"

"Well, I suppose just before Hitler invaded Poland on September 1, 1939, although maybe even as late as September 3 if Hitler had agreed to the British demand that—"

"Excuse me, Professor; that's not what I meant. I mean, when was the last chance that other people could have stopped Hitler from going to war?"

I thought for a while. Throughout the period between the two wars there had been mistakes and miscalculations by statesmen from many countries that contributed to making another war possible. Still, the general consensus is that there was only one European leader actively seeking a new world war, and that was Adolf Hitler.

"That's a tough question," I answered my student. "And my answer is very speculative. But maybe in September 1938, when there was a plot among German generals to take out Hitler just prior to the Munich conference," I said tentatively, trying to recall details.

"I thought the generals' plot took place in 1944, after the Normandy invasion," another student offered.

"Well, there was a military plot in July 1944, led by a Colonel Stauffenberg," I responded. "But there was an earlier plot, less well known, that took place in September 1938, a year before the war actually broke out."

"Well," she asked, "could that plot have succeeded?"

Could it have? And if so, why hadn't it? "You know, Stephanie, that's a good question, but I'm going to have to get back to you on that, okay?"

Thanks to my student's questions, I spent the weekend in the library, tracking down references to the 1938 conspiracy. I found that most historians who had written about it had given it short shrift. Early historians who had lived through and observed the events they described, such as the Englishman J. W. Wheeler-Bennett and the American William L. Shirer, depicted the 1938 conspirators as indecisive, ineffectual, or just plain cowardly. This harsh treatment of the conspirators may well stem from the deep sense of disappointment—even betrayal—that Wheeler-Bennett and Shirer felt when their prewar German acquaintances and friends did not take decisive action against Hitler once his true nature had been confirmed. In any case, while German historians of the same period, Gerhard Ritter and Hans Rothfels, presented a much more sympathetic picture of the conspirators, Shirer's and Wheeler-Bennett's characterization prevailed, certainly among Anglo-American historians.[1]

Second-generation German historians—especially Joachim Fest and Peter Hoffmann—have written cogent defenses of the resistance. Fest argues that the 1938 plot was "probably the most promising of all the plots against Hitler." He concedes, however, that "there are still a number of gaps in the literature on the so-called September plot."[2] Other German historians have been bothered by the apparently treasonable behavior of the members of the resistance during wartime. In particular, the assessment of Hans Oster, a key figure in the resistance from 1938 through 1943, has raised this issue. In 1940, after the war had started, Oster passed on German war plans to the Dutch military attaché. Should he be celebrated for his courageous and eventually redeemed opposition to Nazism, or condemned for divulging secrets that might have led to the deaths of German soldiers? This dilemma has come to be called the "*Osterfrage*," or "Oster question," and the fact that it remains a concern is testimony to Germany's mixed feelings about the resistance.[3]

The most recent book on the German resistance is *On the Road to the Wolf's Lair: German Resistance to Hitler,* by the eminent German American historian Theodore S. Hamerow. His treatment of the 1938 conspirators is the most dismissive of all: "They had no clear plan, no effective organization, little military backing, and even less political influence." Hamerow claims that the conspirators' plan to overthrow Hitler was aborted by the Munich conference, "to the great relief of most of them," a statement that implicitly dredges up the earlier charge of cowardice. Hamerow cites as his witness to the insignificance of the conspirators none other than Sir Nevile Henderson, the British ambassador to Berlin and archproponent of the foreign-policy approach that came to be known as "appeasement." On the subject of the 1938 conspirators, a less credible witness than Henderson—who bent himself into a pretzel to represent British interests and Hitler's goals as nearly identical—can scarcely be imagined. Hamerow's book circles back to the original postwar views of the 1938 conspirators: They were a small band of insignificant, cowardly Nazi collaborators who happily abandoned their jerry-built plot in 1938 when Hitler faced down Chamberlain at Munich.[4]

As I read I became convinced that the history of the 1938 conspiracy had been compromised from the beginning, on both the German and the Anglo-American sides, by personal and political agendas, and by the most deeply held questions of what constitutes patriotic duty.

————

There are two other, more prosaic reasons why historians have not paid much attention to the 1938 conspiracy. First, it never came to fruition. The conspiracy of Colonel Claus von Stauffenberg, who managed to explode a bomb in Hitler's headquarters in July 1944, has received far more attention, despite the fact that it was a failure. Further, even if Hitler had been killed, it would probably not have changed the course of the war, for Germany was already clearly defeated. But in July 1944 at least something happened. A bomb went off, and a handful of officers—primarily in Hitler's East Prussian headquarters, Berlin, and occupied Paris—made a valiant if star-crossed attempt to kill Hitler and overthrow the Nazi regime. In contrast, the 1938 conspiracy has been regarded by most historians as a nonevent unworthy of their time or attention.

Second, until now documents about the 1938 conspiracy have been scarce. Germans conspiring to overthrow Hitler put as little in writing as possible, knowing that if their plot was uncovered they would pay with their lives. In 1944 and 1945, most conspirators did. Only a few of the 1938 conspirators survived the war, and only few of them published formal memoirs detailing their roles in the plot.[5] Although the conspirators generated documents in the form of plans, diaries, and letters, most of these were periodically destroyed; they were simply too dangerous to be kept around. Until very recently what we knew about the conspirators, their plans and their intentions came from the few survivors' memoirs, fragmentary testimony in postwar trials, and traces of the conspirators' visits to London that remain in British archives. As I delved into these sources, I became intrigued by the conspirators' motives, their contacts with the British, and, above all, their plan for seizing power. I wondered, Could they have succeeded?

At an early stage of my research I got lucky. I discovered a valuable cache of documents—the Deutsch Papers in the archives of the U.S. Army Military History Institute at the Army War College in Carlisle, Pennsylvania. The late Harold Deutsch was a distinguished professor of history (and a fluent German speaker) at the University of Minnesota who wrote two relevant books, one about the Blomberg-

Fritsch affair of spring 1938, and the other about the second military conspiracy of 1939–40. He intended to write a book about the 1938 conspiracy but never did.[6]

Beginning immediately after World War II, when he was an interrogator for the OSS, Deutsch established close relationships with the few surviving conspirators, and their widows and other family members. These relationships continued through the 1970s. Deutsch coaxed his contacts into sending him their memoirs, he taped interviews with them, and he maintained an extensive correspondence with many of them. Some of the interviews were transcribed in the 1970s and are available in the archives of the Institut für Zeitgeschichte in Munich. However, most of Deutsch's materials remained in his own files. When Professor Deutsch died in 1995, his children donated his papers to the U.S. Army War College. There they remain today, in twenty-odd boxes, and I am the first historian to have mined these documents for information about the September conspiracy. They shed new light on certain aspects of the conspiracy, enriching our understanding of how it took shape in the late summer and early fall of 1938.[7]

The evidence from the Deutsch Papers and other sources shows that the 1938 conspiracy was well planned and had reasonably good prospects for success. At the center of it stood Lieutenant Colonel Oster, second-in-command of the Abwehr, the Wehrmacht's Office of Military Intelligence. He brought together civilians and military men, resisters of conscience and pragmatic resisters, men with brains and men with weapons, all of whom feared that another war pitting Germany against the combined military and industrial might of the Western Allies would surely end as had the first.

Oster's role has often been misunderstood, in large part because neither he nor his papers survived the war. We know about him only from what others remembered of his actions and his words. Nevertheless, from the Deutsch Papers and other documents, we clearly see Oster eliciting cooperation from the participants and orchestrating plans that involved army officers, policemen, diplomats, and civilians. He was fortuitously placed in the Abwehr, where he had access to many sources of domestic and foreign information, and wide latitude to consult with civilians and officers far above him in the military hierarchy. Finally, because of the covert nature of the Abwehr's activities, Oster had a perfect cover for his conspiratorial work. In 1938 no other lieutenant colonel in the Wehrmacht had a similar freedom of movement.

So, while Hans Oster was one of the true heroes of World War II, he has, for the most part, been either forgotten or reviled by historians. Through 1938 and beyond, Oster resisted Nazi tyranny with guile and cunning, but mostly with great courage. Other key members of the 1938 conspiracy are even less remembered than Oster: Erwin von Witzleben, Erich and Theo Kordt, Susanne Simonis, Hans-Bernd Gisevius, Ernst von Weizsäcker, and Friedrich Wilhelm Heinz. Part of my purpose in writing this book is to ensure that generations to follow will understand that these brave men and women were not cowards or incompetents. Their failure to overthrow Hitler in 1938 was the result not of their lack of will but of the tumultuous state of international relations in September 1938 and the myopia of those who were responsible for gauging Hitler's intentions and responding to his threats.

After three years of research, I can answer my student with confidence: the Oster conspiracy of September 1938 was the last chance for Europeans to stop Hitler from taking the path that would lead to the loss of fifty million lives. Who was responsible for its tragic failure?

Acknowledgments

I owe a great debt of gratitude to Robert and Ingeborg Wolfe. Bob, a former infantryman who still carries a fragment of a Wehrmacht mortar around in his body as a souvenir of World War II, is retired from the staff of the U.S. National Archives after thirty-four years as its senior expert on the Third Reich, the Nuremberg trials, and the U.S. postwar military government of Germany (in which he served for three years). He used his considerable language and research skills to comb through German archives, pulling out important documents. Bob's intimate knowledge of the history of German resistance made him a valuable colleague, with whom I discussed all aspects of the project. Inge, a native speaker of German, did much valuable research and translation, and made crucial suggestions about the readability and accuracy of the text. I could not have written this book without their help. Thanks are also due Margrit Krewson for bringing us together.

Alexandra Weiss also did invaluable translation work, mostly on published German books. My friend John C. Nolan helped in small but important ways.

Professor Peter Hoffmann, of McGill University, Professor Emeritus Russell F. Weigley of Temple University, and Lieutenant Colonel Winfried Heinemann, Ph.D., of the German army, read the entire text and used their knowledge of the German resistance during this period to make many key suggestions. I am grateful to both. Any errors that remain are entirely my responsibility.

Dr. Jeffrey Klepfer, my colleague and friend, provided much-needed institutional assistance from the Office of the Dean of the College of Liberal Arts and Sciences of the University of Tampa, Florida.

I had the help of many librarians and archivists. At the University of Tampa I am indebted to Marlyn Pethe, Elizabeth Lee Barron, Ellie Jones, Jeanne Vince, Keven McGinn, and John Stepro, who all helped in many ways. At the Military History Institute of the U.S. Army War College, Carlisle, Pennsylvania, I drew on the expertise of Dr. Richard Sommers, David Keough, and Jim Baughman. Thanks to Professor Harold Deutsch's children, Janet Leaf, Dorothy Thews, and Harold Deutsch Jr., for giving me permission to quote from the Deutsch Papers.

Gail Winston, my editor, offered superb guidance as I worked to shape the manuscript, and kept the faith throughout. I am also indebted to Gail's assistant, Christine Walsh; to production/copy editor Sue Llewellyn; and to designer Sarah Maya Gubkin for their help in bringing the manuscript to completion.

Above all, I am grateful to my wife, Carol, and my son, Trevor, who gave me unstinting support and encouragement during a particularly difficult period of this book's construction.

Dramatis Personae

(as of September 1938)

THE CONSPIRATORS

Adam, General of Infantry Wilhelm, commander of Germany's "West Wall"

Beck, Colonel General Ludwig, former chief of the Army General Staff

Brauchitsch, Colonel General Walther von, commander in chief of the army

Brockdorff-Ahlefeldt, General Walter, Graf von, commander of Twenty-third Division, Potsdam; subordinate of General Erwin von Witzleben

Canaris, Admiral Wilhelm, chief of the Abwehr (Military Intelligence)

Dohnanyi, Hans von, Ministry of Justice

Gisevius, Hans-Bernd, Department of the Interior

Goerdeler, Carl, former mayor of Leipzig and former Reich price commissioner

Groscurth, Lieutentant Colonel Helmuth, Abwehr

Halder, General of Artillery Franz, chief of the Army General Staff

Hase, Major General Paul von, commander of Fiftieth Infantry Regiment (Landsberg an der Warthe [now Gorzów Wielkopolski, Poland])

Heinz, Captain Friedrich Wilhelm, Abwehr

Helldorf, Wolf, Graf von, police president, Berlin

Hoepner, General of Artillery Erich, commander, First Light Division

Kleist-Schmenzin, Ewald von (the elder), Prussian aristocrat

Kluge, General of Artillery Hans Günther von, commander of Wehrkreis V (Münster)

Kordt, Erich, chief of Ministerial Office, German Foreign Ministry, brother of Theo

Kordt, Theo, counselor, German Embassy, London, brother of Erich

Liedig, Lieutenant Franz Maria, Abwehr

List, General of Infantry Wilhelm, commander, Wehrkreis IV (Dresden)

Nebe, Arthur, chief of Criminal Police (Kripo)

Olbricht, Lieutenant General Friedrich, chief of staff to General List, Wehrkreis IV (Dresden)

Oster, Lieutentant Colonel Hans, Abwehr, organizer of the conspiracy

Röhricht, Lieutenant Colonel Edgar, staff officer to Generalleutnant Olbricht

Schacht, Hjalmar, head of the Reichsbank, former minister of finance

Schlabrendorff, Fabian von, lawyer; historian of the resistance movement; not directly involved in the 1938 plot

Schulenburg, Fritz-Dietlof, Graf von der, vice president, Berlin Police

Simonis, Susanne, reporter for the *Deutsche Allgemeine Zeitung*, cousin of Erich and Theo Kordt

Strünck, Theodor, and Elisabeth Gärtner-Strünck, friends of Oster's and sympathizers with the resistance

Stülpnagel, Lieutenant General Carl-Heinrich von, deputy chief of staff II on Army General Staff

Ulex, General of Artillery Alexander, commander of XI Army Corps (Hannover)

Weizsäcker, Ernst von, state secretary, German Foreign Ministry

Wiedemann, Captain Fritz, Hitler's adjutant

Witzleben, General of Infantry Erwin von, commander of Wehrkreis III (Berlin)

Witzleben, Lieutenant Colonel Hermann von, cousin of Erwin von Witzleben

The Nazis

Goebbels, Joseph, minister of propaganda

Göring, Reich Marshal Hermann, commander of the Luftwaffe and Hitler's designated successor

Heydrich, Reinhard, commander of the SD (Sicherheitsdienst—Security Service) and the Gestapo

Himmler, Reichsführer-SS Heinrich, commander of the SS

Hitler, Adolf, chancellor and führer

Kaltenbrunner, Ernst, SD subordinate and, after 1942, successor of Heydrich

Lorenz, Werner, SD official

Ribbentrop, Joachim von, foreign minister

The British

Cadogan, Sir Alexander, permanent undersecretary, Foreign Office

Chamberlain, Neville, prime minister

Churchill, Winston S., Conservative member of Parliament

Dawson, Sir Geoffrey, editor in chief, *The Times*

Duff Cooper, Alfred, First Lord of the Admiralty

Halifax, Edward Wood, third Viscount Halifax, foreign secretary

Henderson, Sir Nevile, ambassador to Germany

Kirkpatrick, Sir Ivone, first secretary at British Embassy, Berlin

Simon, Sir John, chancellor of the exchequer

Stanley, Oliver, president of the Board of Trade

Vansittart, Sir Robert, chief diplomatic adviser to the government

Wilson, Sir Horace, chief industrial adviser to the government, confidant of prime minister

OTHERS

Beneš, Edvard, president of Czechoslovakia

Blomberg, Field Marshal Werner von, former war minister

Daladier, Edouard, prime minister of France

Engel, Captain Gerhard, Hitler's army adjutant

Frank, Karl Hermann, official of Sudeten German Party

Fritsch, Colonel General Werner von, former commander of the army

Henlein, Konrad, leader of the Sudeten German Party

Hossbach, Colonel Friedrich, Hitler's former adjutant

Jodl, Major General Alfred, chief of Operations Office, Oberkommando der Wehrmacht (OKW)

Keitel, General of Artillery Wilhelm, chief of OKW; in fact, Hitler's administrator

Liebmann, General of Infantry Curt, Fifth Army commander

Masaryk, Jan, Czech ambassador to Great Britain

Raeder, Admiral Erich, commander of the navy

Schmidt, Paul Otto, Hitler's interpreter

Prologue

On the afternoon of September 5, 1938, as Europe seemed about to be thrown into a war more brutal and more terrifying than any it had ever known, a young woman named Susanne Simonis boarded a flight from Berlin's Tempelhof Airport bound for London, ostensibly pursuing a story as a foreign correspondent for the *Deutsche Allgemeine Zeitung*. In fact she was acting as a messenger between her cousins, two brothers who worked for the German Foreign Ministry: Erich Kordt, deputy to Foreign Minister Joachim von Ribbentrop in Berlin, and Theo Kordt, counselor at the German Embassy in London. It was a mission Simonis had undertaken several times in past months. Not wanting to risk a search at customs, she had memorized the message that Erich had given her in Berlin, and would recite it to Theo in London.

Of course, as members of Germany's diplomatic corps, the Kordt brothers need not have been concerned about the scrutiny of customs clerks. But they had good reason to forgo the diplomatic pouch and to use a trusted family member as a courier between them. They, along with Lieutentant Colonel Hans Oster of the Abwehr (Military Intelligence), were at the center of a conspiracy to seize power and overthrow Adolf Hitler and his Nazi regime.

Over the past four years Hitler had been tearing up the Versailles Treaty, provision by provision, daring France and Britain to stand up to him. Now, in September 1938, he was demanding that Czechoslovakia cede a part of its territory with a sizable German population—the so-called Sudetenland—to Germany. If the Czech government refused to comply with his demand, Hitler threatened that Germany would take things into its own hands and launch an attack to "free" the Sudeten Germans from Czech rule. Such an attack on a sovereign nation with treaty ties to the Soviet Union and France, and through France, to Great Britain, would surely lead to a second world war on the continent of Europe, this one promising to be even more destructive than World War I, which had left millions dead or maimed.

The imminent prospect of such a new war accelerated the plans of a number of German military officers and civilians to overthrow the Nazi regime. This developing conspiracy brought men and women who objected to Hitler on moral grounds into league with officers in the German army who were drawn into the conspiracy because they feared that the "Austrian corporal's" raging thirst for war would bring another humiliating defeat to the German military and ruin to their beloved fatherland.

The governing spirit of the Nazi regime was embodied in one man—Adolf Hitler. Although most of the conspirators wanted only to arrest Hitler and put him on trial, Hans Oster and a few others had decided to seize the centers of power in Berlin and kill Hitler in the Reich Chancellery before his supporters could organize a countercoup and restore him to power. All they needed before taking action was assurance from the British that they would not give in to Hitler's newest territorial demands and make him even more of a national hero than he had already become.

Once she had landed in London, Susanne Simonis repeated the memorized message from Erich Kordt to his older brother, Theo. It was then Theo's turn to take the message to Number 10 Downing Street, where he had secured an appointment with Sir Horace Wilson. Like many English homes, Number 10 has a back garden. It was through this garden gate, shielded from prying eyes, that Theo Kordt entered on September 6.

Although Sir Horace Wilson held the innocuous-sounding title of chief industrial adviser to His Majesty's Government, his real role was that of confidant and right-hand man to Prime Minister Neville Chamberlain. Sir Horace, an ardent proponent of appeasement at almost any price, was not eager to be drawn into a conspiracy to overthrow the legitimate government of Germany. Nevertheless, since late July he had been meeting regularly with Theo Kordt, and was sufficiently intrigued by his latest message to invite him to return the next day, again via the back garden gate, to meet secretly in his office with the foreign secretary, Lord Halifax. Here was the kind of opportunity that Kordt had been hoping for—a chance to make the conspirators' case to the second-highest-ranking member of the British cabinet. Kordt knew that on this conversation rested the fate of the conspiracy, the fate of Germany, and ultimately, the fate of Europe.[1]

1

The Führer Makes War on His Army:
Oster Recruits the Generals

Adolf Hitler had come to power on January 30, 1933, as the chancellor of a coalition government. Within months he and his Nazi
henchmen had ruthlessly seized total power, crushed all political opposition, and blatantly violated the rule of law.

Some Germans had applauded these measures; others had accepted
them passively; a few had been repelled. One in the last category was
Hans Oster, the son of a pastor and a career military officer who had
served with distinction in World War I. Oster had chosen to remain in
the diminished Reichswehr (German armed forces) when it had
been reduced to a mere one hundred thousand men by the Treaty of
Versailles, and had served in a variety of postings throughout the
1920s.

Oster had found the Nazis' politicization of all aspects of German
life "deeply unpleasant," but, like many Germans, he had cautiously
approved of their politics of national strength. He had kept his own

counsel and watched as Hitler secretly began to rebuild the German military machine.

Oster, a gregarious and charismatic man, made friends easily and inspired the trust of others. By early February 1934, he had won a position in the Abwehr, the German Office of Military Intelligence. A few months earlier Oster had met Hans-Bernd Gisevius, an ambitious young man who had taken a position in the Nazis' new instrument of terror, the Gestapo, only to be alarmed by what he had learned. As Gisevius began to show his new friend secret papers from Gestapo headquarters, Oster had been shocked by their contents. It was clear that what the Nazis had already done—such as imprisoning political dissidents and religious figures and setting up the first concentration camps—was a mere prelude to what was coming. He began to agree with Gisevius that the Nazi regime was "a robber's cave." The enlightenment of Hans Oster had begun.

But it was not until Hitler's massacre of hundreds of SA (Sturmabteilung) leaders on June 30, 1934, "the Night of the Long Knives," when two senior army generals were killed, that Oster became convinced that the regime was evil and must be overthrown. The slaughter of June 1934 was, Oster later reflected, the first opportunity to "nip the methods of these robbers in the bud," and the army had not only failed but, by providing arms and transport to the black-shirted SS murderers, had been complicit in the lawlessness of that event. In the year and a half since Hitler had come to power, he had made war inexorably on all Germans who dared question his authority. Oster was converted from a skeptic about the regime into a resister.[1] What pushed him still farther along the path toward assassination was Hitler's treatment of the Jews.

Jews had suffered occasional street violence and even murder in the tumultuous years of the early 1920s, but when the Nazis took power, these incidents had escalated. Policemen had stood idly by as SA hoodlums smashed the windows of shops owned by Jews, beat Jewish boys and men, and humiliated Jewish women. Signs had begun to appear in public places, saying THIS AREA IS JEW-FREE! or DOGS AND JEWS NOT ALLOWED! Germans had been encouraged to sever relationships with Jews. The carriers of "Jewish infection" were to be isolated from "Aryan" society. In 1935 the Nazis had codified this policy in the Nuremberg laws, which, among other provisions, stripped Jews of their citizenship, forbade intermarriage between Jews and "pure Germans," and had made it illegal for Jews to employ gentile servants.[2] Between 1933 and 1938, as the Nazis turned the screws more tightly, 150,000 Jews—about 30 percent of Germany's total—had emigrated.[3]

One of Oster's closest friends was Erich Schultze, who recalled that "Oster and I had complete trust in each other during the Third Reich and were able to talk openly. From early on there was no doubt in our minds about the necessity of Hitler's removal; only how was left unanswered."[4] The two did not suddenly awaken to the realization that Hitler must be removed. According to Schultze they "didn't grow into this task as soldiers or officers but rather as conscientious and concerned Christians who were very worried about the increasing danger facing the entire world." The most important motive behind their growing hatred for Nazism was concern about Germany's Jews, "who were being driven to their destruction." Schultze wrote, "We suddenly felt we were responsible before God for their [the Jews'] rescue." By 1937, according to Schultze, Oster had decided that Hitler must be killed.

Oster had been moved by moral opposition to Hitler's policies. But many of his brother officers, who were indifferent to Hitler's domestic savagery, were alarmed by Hitler's reckless foreign policy, which they feared would lead Germany into a suicidal war. Between 1933 and 1936 the führer had taken Germany out of the League of Nations, announced the existence of an air force (Luftwaffe) and revealed plans for conscription that would lead to a half-million-man army, and remilitarized the Rhineland. These last actions were unilateral violations of the Treaty of Versailles, which had ended World War I. While Hitler had dismantled the treaty, his former enemies, Great Britain and France, had registered only feeble protests. But surely, the generals thought, Hitler would eventually push the British and French too far, and then there would be war.

Between 1935 and 1937 Oster had made contact with like-minded men who would eventually become his allies in 1938. But during that period there had been no opportunity for him to take action against the hated Nazi regime. As long as Hitler had been able to get away with bullying the Western powers, most Germans had shrugged off whatever doubts they may have had about their führer. After all, he had brought prosperity to their depression-ridden country and had restored the fatherland to its rightful place as a strong military power in Europe.

Then, in November 1937, two meetings took place that changed the course of European history. These secret meetings set off a chain of events that brought Germany to the brink of war, and created a heaven-sent opportunity for Hans Oster to set in motion his plan to eliminate Hitler and overthrow the Nazi regime.

November 5, 1937, Berlin, Reich Chancellery

In the fading light of the dank afternoon of November 5, 1937, a fleet of sleek Mercedes-Benz sedans pulled into the entrance of the Reich Chancellery at the intersection of Wilhelmstrasse and Voss Strasse and disgorged their famous passengers. One by one they entered the building, receiving the salutes of the black-uniformed SS guards lining the entrance and passageways. The four-story neo-Renaissance building, constructed in the mid–nineteenth century, had been the residence and workplace of German chancellors since the time of Prince Otto von Bismarck, the architect of German unification. It remained the heart of government, but by 1937 the tricolor flag of the Weimar Republic— black, red, and gold—no longer flew over the building. It had been replaced by the old imperial red, black, and white with a superimposed Nazi swastika. Accompanied by their retinues, the princes of the Third Reich made their way to the conference room but entered alone, leaving their adjutants and technical experts to wait in the anteroom. They had been summoned to a conference by Adolf Hitler.

Gathered around the heavy oak table were the men who commanded the armed forces of the Reich: Colonel General Hermann Göring, commander of the air force; Admiral Erich Raeder, commander of the navy; Colonel General Werner von Fritsch, commander

Government District of Berlin, 1938

DOROTHEEN STR.

HEERESBÜCHEREI

FRANZÖSISCHE BOTSCHAFT

REICHS-MIN. FÜR DES INNEREN

BRANDENBURGER TOR

UNTER DEN LINDEN

PARISER PLATZ

HOTEL ADLON

REICHS-MIN. FÜR WISSENSCHAFT ERZIEHUNG u. VOLKSBILDUNG

AKADEMIE D. KÜNSTE

BEHREN STRASSE

ENGLISCHE BOTSCHAFT

MAUER STR.

GENERAL-JNSP. FÜR D DEUTSCHE STRASSEN WESEN

REICHS-MIN. FÜR ENÄHRUNG u. LANDWIRTSCHAFT

WILHELM STRASSE

REICHS-JUSTIZ-MIN.

REICHSPRÄSIDENTEN-PALAIS

AUSWÄRTIGES AMT

PREUSS. STAATSRAT u. VERBINDUNGS STAB DES STELLVERTRETERS DES FÜHRERS

REICHSKANZLER-PALAIS
(HITLER'S RESIDENCE)

REICHS-PROPAGANDA-MIN. REICHSKULTUR KAMMER

REICHSKANZLEI
(REICH CHANCELLERY)

KANZLEI DES FÜHRERS

WILHELM-PLATZ

HOTEL KAISERHOF

GAULEITUNG GR.-BERLIN DER NSDAP

VOSS STR.

GENERAL-DIR. DER REICHSBAHN

REICHS-FINANZ-MIN.

REICHS-VERKEHRS-MIN.

STRASSE

LEIPZIGER

REICHS-POST-MIN.

PREUSS. STAATS-MIN.

PRIVAT KANZLEI DES FÜHRERS

REICHS-LUFTFAHRT-MIN.

ZIMMER STRASSE

HAUS DER FLIEGER

PRINZ ALBRECHT STRASSE

GEHEIME STAATSPOLIZEI
(GESTAPO HEADQUARTERS)

Note Reich Chancellery (78 Wilhelmstrasse), Hilter's adjacent residence, and Gestapo Headquarters on Prinz Albrecht Strasse.
Source: *Topography of Terror—Gestapo, SS, Reichssicherheitshauptamt on the "Prince-Albrecht-Terrain": A Documentation,* edited by Reinhard Rürup (Berlin: Topography of Terror Foundation, 1987), page 15.

of the army; and Field Marshal Werner von Blomberg, war minister and commander in chief of the Wehrmacht. Also at the table sat Baron Konstantin von Neurath, foreign minister, and Colonel Friedrich Hossbach, Hitler's military adjutant, whose minutes of this historic meeting came to be known as "The Hossbach Memorandum."[5] Their pathways to eminence in the German establishment had been straightforward and orthodox, with the sole exception of Göring—only he owed his position to his Nazi connections.

In his youth Hermann Göring had been one of the German military's bright stars. In October 1914, at the age of twenty-one, and just weeks after the outbreak of World War I, he had been chosen to become one of the fatherland's first fighter pilots. He was fearless and successful, winning Germany's highest military decoration, *Pour le Mérite*, and ending the war as commander of the famed Richthofen Squadron. He joined the embryonic Nazi Party in 1922, and was wounded in the abortive "Beer Hall Putsch" the next year.

In contrast to other high-ranking Nazis, Göring was an affable and even charming man who usually made a favorable impression on foreign visitors. He was flamboyant and self-indulgent, with an inordinate love of luxurious homes, gaudy uniforms, and good food. Indeed, his weakness for the last earned him the nickname of *der Dicke*, "the Fat One." He was one of the very few top-echelon Nazis to whom ordinary Germans felt connected, perhaps because of his visible human weaknesses. From the early 1930s until 1942, Göring was effectively number two in the Nazi hierarchy.[6]

Göring was publicly acknowledged to be the commander in chief of the Luftwaffe in 1935. By November 1937, though the Luftwaffe had grown considerably, it was far from being the dominant force it became two years later. Ominously, the weapons of the blitzkrieg—the Messerschmitt fighters, the Heinkel medium bombers, and the Stuka dive-bombers—which would soon terrorize all of Europe, were being combat-tested by German pilots fighting alongside the fascist forces of General Franco in Spain's civil war.

Erich Raeder, at sixty-one the oldest man in the room, had been appointed commander in chief of the navy in 1928, five years before

Hitler's ascent to power. Although he was a career naval officer who had climbed steadily through the ranks, Raeder showed none of the reserve—even contempt—that regular army officers had for Hitler. Indeed, Raeder had embraced Hitler from the start.

Hitler, as a former infantryman, had little understanding or interest in naval matters. That was a mixed blessing for Raeder. While it meant that Hitler almost never meddled in naval affairs, it also meant that he gave Raeder only minimal support. In 1937 the navy was in the midst of a building program that was scheduled to reach maturity in 1947. But in 1937 Raeder's capital ships included only three pocket battleships, two heavy cruisers, and two battle cruisers—not a match for even the French navy, and far inferior to the Royal Navy. Of course, in the event of war, the Royal Navy, with its global commitments, would leave only a fraction of its strength in home waters. But Raeder was well aware that his puny force would be blown out of the water in a showdown with Britain, even with a reduced fleet.

Werner von Fritsch was the main architect of the renascent German army. In 1937 that job was far from over. The army numbered just over forty divisions, including eight motorized and three panzer divisions. It was far inferior to the French army, which could field nearly one hundred divisions. Although the French armor was not formed into separate divisions, French armored brigades included tanks that were heavier than their German counterparts. Fritsch contemplated the idea of war with horror, and especially a war with France, Britain, and the Soviet Union. The army was years away from a strength that would make such a conflict even thinkable.

Werner von Blomberg had served as staff officer in World War I, and continued to serve as a staff officer in the 1920s.[7] In 1933 he was appointed minister of war in Hitler's first cabinet. Blomberg served Hitler ably but was never a Nazi insider.

Konstantin von Neurath was a career diplomat who had been appointed foreign minister in 1932, the last troubled year of the Weimar Republic. Hitler kept him in office largely to assure the world that the Nazis intended to initiate no great changes in foreign affairs. Neurath was a compliant character who kept his position only so long as he was able to suppress his better judgment and support Hitler's adventurist foreign policy.

Hitler addressed his remarks to these men that November afternoon. He had come to speak, not to consult. Hitler spoke with force and emotion, using his clenched right fist, forefinger protruding like a revolver aimed at his listeners. He leaned toward them and then drew back, occasionally thumping the table to accentuate a point. Those who were subjected to the führer's private oratory found it unnerving, and were often at a loss for words; women sometimes swooned; men were intimidated or beguiled.

The meeting began at 4:15 and concluded at 8:30, with Hitler holding forth for the first two hours. He started by reiterating the theme of his book, *Mein Kampf*, published a decade earlier. Germany needed *Lebensraum* ("living space") for the German race to expand and prosper. This would not be found in colonies in Africa or Asia but in Europe, "in immediate proximity to the Reich." But for Germany to seize such space, it must be prepared for conflict, for "there had never . . . been spaces without a master. . . . [And] the attacker always comes up against a possessor." Standing in Germany's way, Hitler claimed, were "hate-inspired" Britain and France. Nevertheless, both were presently weakened and unlikely to respond to a German offensive, Britain because of its troubles in Ireland and India, and France because of "internal political difficulties."

Under any circumstances, Hitler said, Germany needed to strike no later than 1943–45, for by then "our relative strength would decrease in relation to rearmament . . . by the rest of the world." The führer then outlined several political scenarios that might present Germany with an opportunity for action. The thrust of his message augured a much earlier move against his enemies. He foresaw the possibility that "if internal strife in France should develop into such a crisis as to absorb the French Army completely and render it incapable of use for war against Germany, then the time for action against the Czechs had come." Or France might be embroiled in war with "another country" (that is, Italy), which would also render it vulnerable. In either case, Hitler said, there could be no hesitation about *where* Germany would strike: "Our first objective must be to overthrow Czechoslovakia and Austria simultaneously to remove the threat to our flank in any possible operation against the West."

So the targets were to be the two states created by the Versailles peacemakers to keep Germany contained in central Europe. Austria, once the heart of the Hapsburg Empire, had now been reduced to a mountain republic of seven million German-speaking citizens, forbidden from ever joining with Germany. Czechoslovakia, the great democracy of eastern Europe, was a multiethnic state (including 3.5 million Germans) with strong mountain defenses and the Skoda munitions works, the most impressive in Europe. Both Austria and Czechoslovakia blocked Germany on its southern and southeastern borders. With these two states annexed, Hitler claimed, the Reich would expand to its "natural" frontiers, and could produce more foodstuffs and raise another twelve divisions for the Wehrmacht.

But what of Britain and France? The latter had guaranteed the Czech borders through a military assistance treaty. While Britain was not party to that treaty, it was France's strongest ally, and no one could envision a war between France and Germany in which Britain would not become involved on the French side. But Hitler shrugged off this threat, believing that "almost certainly Britain, and probably France, had already written off the Czechs." Even if the French wanted to defend Czechoslovakia, he predicted, they would not do so without British support, which would never be forthcoming.

Although it was impossible to foresee precisely when an opportunity might present itself, Hitler was determined to take advantage of it "even as early as 1938." And when the time came, Hitler said, "the descent upon the Czechs would have to be carried out at lightning speed." Hitler had spelled out his plans. War was inevitable. It could come at any time between 1938 and 1943–45, but the commanders must be ready to act whenever political events were favorable. The war machine he had been preparing for the last four years must be ready for use as early as two months thence.

After his monologue, Hitler asked his chiefs to respond. Raeder was silent, but one can guess his inner thoughts as he contemplated fighting the Royal Navy with his handful of warships. Fritsch and Blomberg were not so reticent. They questioned Hitler's assumption of French vulnerability. Even if France were engaged in war with Italy, its army was still sufficient to bring overwhelming force to bear against the much smaller German army. Blomberg noted the weakness of German fortifications in the west and contrasted these to the "strength of the Czech fortifications which had now acquired a structure like a Maginot Line which would gravely hamper an [German] attack."[8] Both army men expressed horror at the possibility of being at war with both France and Britain. Neurath questioned Hitler's assertion that an Anglo-French-Italian conflict was likely soon.

Göring spoke only when he was attacked by Blomberg and Fritsch for his mismanagement of the Four-Year Plan, which would have to supply the matériel that the generals needed to fight a war. Göring had assumed responsibility for the plan in 1936, but he had little interest or competence in economic and planning matters. As Göring endured a merciless tongue-lashing by the two army men, Hitler was "an attentive listener." For this humiliation the Luftwaffe chief would soon find a way to strike back at his antagonists. For his part Hitler left the meeting disgusted by his army commanders. He had ordered them to prepare for war, and they had had the temerity to oppose him. There would be war with them or without them.[9]

Fritsch immediately told Colonel General Ludwig Beck, chief of the Army General Staff, what Hitler had said. Since the Abwehr offices were located in the Army Headquarters Compound on Bendlerstrasse, and since Fritsch was close to Oster and Admiral Wilhelm Canaris, director of the Abwehr, it is likely that both Abwehr officers also became privy to Hitler's plans at this time. Like Fritsch and Blomberg, their first reaction was surely horror at the thought of another war. But Oster may also have realized that Hitler's blatant aggression against Austria and Czechoslovakia would reveal to the German people that their führer was a megalomaniac determined to drag them into war.

Just two weeks after announcing to his military commanders that he intended to take over Austria and Czechoslovakia before his military advantage over Britain and France was lost, Adolf Hitler came face-to-face with a member of the British ruling class. The meeting was an opportunity for the führer and a representative of Neville Chamberlain's new government to take each other's measure.

Edward Wood, the third Viscount Halifax, had come to Germany for the hunting. At least that's the way the story went. Halifax loved field sports, especially foxhunting and shooting. Born without a left hand, he used a prosthesis in the shape of a clenched fist with a spring-activated "thumb" to open field gates while holding the reins and horn in his right hand. Extremely tall and gaunt, he gave the impression of great solemnity. Among his contemporaries there are no recorded Halifax bons mots, and few recollections of hearty laughter. Like many British aristocrats, he spoke with a slight lisp and developed the habit of extreme circumlocution, which even his friends occasionally found baffling. He once began a sentence, "I should have thought that one might say that it could be reasonably held that . . ." Today we might call him shy, even asocial, but in the small world of London clubs, country house parties, foxhunts, and Tory politics, he was respected and admired. He maintained a professional cordiality but personal reserve with all but his very few intimate friends. After his sons married, his new daughters-in-law asked what they should call him. "Lord Halifax," he replied.[10]

Halifax's wedding to Lady Dorothy Onslow in 1909 marked the beginning of a long and happy marriage that produced several children and no scandals. He entered politics in January 1910, winning the Tory stronghold of Ripon, Yorkshire, near his family seat. Another poll eleven months later was his last contested election in a political career that lasted nearly fifty years. Halifax made steady progress up the political ladder. By September 1938, he had served as president of the Board of Education, viceroy of India, secretary for war, Lord Privy Seal and, from February 1938, foreign secretary. He was regarded by his peers and by other politicians as a man of sound principles and good judgment.

In November 1937 Lord Halifax, as a private person, in his capacity of master of the Middleton Hunt, had been invited to the International Hunting Exposition in Berlin by another private person, Hermann Göring, in *his* capacity as *Reichsjägermeister*, "Reich master of the hunt." Göring intimated that, if Lord Halifax wished, an interview might be arranged with the führer. The hunting trip charade was necessary because neither government wanted to risk the embarrassment of an official meeting that might turn out badly, but each wanted to make contact with the other.

In 1937 Lord Halifax held the cabinet position of Lord President of the Council. More important, he was an ally and confidant of the British prime minister, Neville Chamberlain, with whom he shared a commitment to the policy of appeasement. After the outbreak of war in September 1939, this policy would be castigated as a monumental failure, but in 1937 it had wide public support and was regarded as generous and progressive. In Chamberlain's view appeasement was recognition of Germany's de facto return to great-power status. Chamberlain tacitly admitted the injustice of the Versailles treaty by his willingness to allow Germany to make adjustments to it, including the restoration of some of the colonies and European territories the treaty had taken away. Chamberlain also assumed that once these adjustments were made, Hitler's appetite for territory would be sated, and Germany would settle in and conduct itself like a responsible great power.

In return for making concessions to Hitler, however, Chamberlain wanted a "general settlement" between the two countries. He expected German recognition of the newly altered boundaries, and a pledge that Britain and Germany would peacefully share leadership in Europe. The pledge did not have to be as formal as a treaty, but it did need to be fully discussed, agreed upon, and signed. Chamberlain's quest for this elusive general settlement would drive his diplomacy toward Hitler from November 1937 through the Munich conference eleven months later.

When Göring's invitation arrived in London in early October,

Anthony Eden, the British foreign secretary, had serious reservations about whether Halifax should accept it. Eden, who was not a proponent of appeasement, and increasingly at odds with Chamberlain, feared that Halifax would deliver the wrong message. Chamberlain, however, thought that the invitation was a wonderful opportunity to raise the idea of a general settlement with Hitler, and Halifax himself was eager to go. Thus on November 17, 1937, Halifax flew to Berlin and was lavishly feted by German hunting enthusiasts. (Indeed, they nicknamed him "Halalifax," a play on the cry *Halali!* the German equivalent of British foxhunters' "Tally-ho!") Halifax was frankly attracted by Göring's "composite personality . . . film star, great landowner interested in his estate, Prime Minister, party-manager, head gamekeeper at Chatsworth."[11] Two days later, his hunting charade over, Halifax took the train to Bavaria to meet Hitler at his mountain hideaway.

The Bavarian town of Berchtesgaden is nestled beneath the looming mountain of Obersalzberg on the German-Austrian border. Hitler had been coming to the area since 1925, and had purchased a small house and property on the mountain in 1935. With his growing personal wealth, derived from royalties from *Mein Kampf* and money paid to him by the German government for using his picture on postage stamps, Hitler had been able to expand both house and property. By 1938 his holdings included nearly eighteen hundred acres and a greatly enlarged house, "the Berghof," with an enormous picture window overlooking the town of Berchtesgaden and Untersberg Mountain. Hitler said, "By night at the Berghof I often remain for hours with my eyes open, contemplating from my bed the mountains lit up by the moon." Hitler spent as much time as he could there, often conducting business from this remote location.[12]

After a ride up the mountain, Halifax narrowly averted an early disaster as he was getting out of his car. Hitler was decked out in local costume, which included "black trousers, white silk socks, and pumps." Halifax assumed that he was a footman, and was about to hand him his hat and coat when Neurath, the German foreign minister, whispered hoarsely, "*Der Führer! Der Führer!*" Halifax barely avoided mistaking the dictator of one of the world's most powerful military powers for a servant in livery.

The two finally settled into the deep, oversize easy chairs that Hitler favored, the six-foot eight-inch, horse-faced, impassive Halifax facing the short, mustachioed bundle of twitches wrapped in Bavarian garb. Hitler began the discussion with an impassioned diatribe about the impossibility of dealing with democracies. Then, in a shrewd bit of misdirection, the führer claimed that "between England and Germany there was only one difference, the colonial issue." In fact, Hitler had little interest in colonies. Nevertheless Halifax seized on the colonial issue to raise the possibility of "a genuine settlement by means of which quiet and security might be established in Europe." Hitler said nothing. Halifax went further, expressing hope that Germany might rejoin the League of Nations, which it had left four years earlier. Still no response from Hitler. Finally Halifax played his trump card. According to his notes on the meeting, he did precisely what Eden had feared he would: He gave Hitler the green light to further aggression. Halifax knew that Hitler coveted territories in Central Europe that were inhabited by ethnic Germans but had been denied to Germany by the Treaty of Versailles. Specifically these included Danzig (now Gdansk), the German city on the Baltic, which had been made into a "free territory"; the Sudetenland on Germany's southeastern border, which was given to Czechoslovakia; and the independent republic of Austria, which was forbidden from ever joining with Germany. Halifax's notes faithfully summarize his representation of Britain's position:

> *All other questions fall into the category of possible alterations in the European order, which might be destined to come with the passage of time. Amongst those questions were Danzig, Austria, and Czechoslovakia. England was interested to see peaceful evolution and methods should be avoided which might cause far-reaching disturbances, which neither the chancellor nor other countries desired.*[13]

In this astounding statement a senior member of the British government had unmistakably signaled Hitler that he was free to annex

Austria and Danzig and to dismember Czechoslovakia, as long as he did it in a gentlemanly fashion.

Hitler finally answered that "as far as Czechoslovakia and Austria were concerned, a settlement could be reached given a reasonable attitude." He went on to make clear that this "reasonable attitude" would have to come from the Austrians and the Czechs. The session concluded with Hitler again raising the issue of the return of German colonies, no doubt to camouflage his glee at having gotten what he wanted without having had to ask for it.

At the vegetarian lunch that followed, conversation turned to India, where Halifax had served as viceroy from 1926 to 1931, during a period of nonviolent agitation by Indian nationalists. Hitler offered his guest some advice about how the British should handle these trouble-makers. "Shoot Gandhi," he said, "and if that does not suffice to reduce them to submission, shoot a dozen leading members of [the] Congress [Party]; and if that does not suffice, shoot 200 and so on until order is established."[14]

Halifax gazed at Hitler "with a mixture of astonishment, repugnance, and compassion," according to Ivone Kirkpatrick, but maintained his composure.[15] Oddly enough, that was not the last time Gandhi's name came up. When he returned to England, Halifax several times compared Hitler to the Indian nationalist, once describing Hitler as being "like Gandhi in Prussian boots." The only way to make sense of this bizarre comparison is that, from Halifax's perspective, both were troublesome, strange-looking nationalists who refused to negotiate according to conventional rules. This misconception serves as a reminder of the myopia of British appeasers: In Hitler they thought they had an eccentric but basically reasonable fellow with whom they could "do business."

When he returned to Berlin after his interview with Hitler, Halifax was greeted by his host, the Reich master of the hunt, "dressed in brown breeches and boots all in one, green leather jerkin surrounded with a green leather belt, on which was hung a dagger in a red leather sheath." Despite Göring's buffoonish appearance and genial bonhomie, Halifax wondered "how many people he had been responsible for getting killed."[16]

That evening at dinner Halifax sat next to War Minister Blomberg, who, in contrast to Hitler, told him the truth: Germany cared little about colonies; the "real issues" were in central and eastern Europe. On his return to London by train and boat, on November 22, Halifax mused about the differing messages that he had received and the gulf in values that separated him from the German dictator. Nevertheless he assured himself that at least he had "made contact," and that the visit had done nothing to harm Anglo-German relations.[17]

Eden, when he saw Halifax's notes of the meeting with Hitler, thought otherwise. In his indecent rush to concede Germany freedom of action in central Europe, Halifax had given away too much. Despite Hitler's seeming indifference, Eden feared that the führer might have interpreted Halifax's remarks as a signal from His Majesty's Government that he was free to pursue his heart's desire without worrying about British interference. His suspicions were well founded. Shortly after Halifax's departure, German Foreign Minister Neurath sent a secret telegram to German embassies in Rome, London, Paris, and Washington, summarizing the talks.

> The British did not believe that the status quo had to be maintained under all circumstances. Among the questions in which changes would probably be made sooner or later were Danzig, Austria, and Czechoslovakia. England was only interested in seeing that such changes were brought about by peaceful development.[18]

In London, Chamberlain expressed delight on hearing the details of Halifax's mission. In the prime minister's opinion Halifax had made a very successful initial contact with the German dictator. Back at the Berghof, Hitler contemptuously dismissed Halifax as "the English parson."[19] He was reassured that he had been correct at his meeting with his military commanders on November 5 when he had predicted that "Britain has already written off the Czechs."[20] Now satisfied that his instinct about Britain had been correct, Hitler could begin to accelerate his plans for conquest in central Europe.

JANUARY 12, 1938, BERLIN, WAR MINISTRY

The cozy group in the war ministry included the War Minister himself, Field Marshal Blomberg, his three current and one former adjutants, a pretty young woman and her mother, Göring, and a fidgety führer. The latter two had come to serve as witnesses at the wedding of Germany's leading soldier.[21] Blomberg was a tall, handsome, outgoing, sixty-year-old who had been widowed since 1929. In the fall of 1937, Blomberg had fallen in love with Eva Gruhn, a twenty-five-year-old whom he had met at "the White Hart Hotel," where she worked. Fräulein Gruhn was, in the words of her lover, "a simple child of the people." Her father, a gardener, was deceased, and her mother was the proprietress of a massage parlor of uncertain reputation. However enamored he was of the young woman, Blomberg appreciated that the officer corps and perhaps the führer might not approve of his marriage to a woman so young and so lacking in social standing. However, circumstances forced Blomberg's hand by mid-December. The field marshal had a rival: a young man who also sued for Eva's favor and who refused to disappear. Furthermore, Fräulein Gruhn was several months pregnant. Blomberg decided to marry her.

Blomberg's next move, in retrospect, appears to be an action of monumental self-destructiveness. He confided his dilemma to Hermann Göring and asked for his help. He had fallen in love and was having an affair with a young woman "with a certain past." How did Göring think Hitler would react to the general's desire to marry Fräulein Gruhn? And could Göring do something about his inconvenient rival?

Göring immediately recognized Blomberg's request as a rare opportunity to engineer the general into disgrace and out of office, an office that Göring himself coveted. *Of course* he would help. Within days the young rival received an attractive offer of employment with a German firm in Argentina, which he accepted with alacrity. Göring assured Blomberg that he would have a word with the führer about the marriage. Soon the path had been smoothed out. Hitler not only approved of the marriage, but he, along with Göring, agreed to serve as a witness. The relieved field marshal quickly consummated his nuptial plans before the führer could change his mind and the bride's condition became too obvious.

The trap was not sprung until Blomberg and his bride left on their honeymoon. Then a file on the new Frau Blomberg suddenly appeared on the desk of a Berlin police official. In the recent past she had been arrested, albeit never convicted, on morals charges. Although Hitler and others later referred to her as a prostitute, the only offense of which she was apparently guilty was appearing in pornographic photographs.

Eventually the fortuitous police file made its way up the hierarchy and landed on Göring's desk, as he surely intended it would. Bemoaning his fate always to be the bearer of bad news to the führer, Göring presented the material to him on January 24. Hitler appeared genuinely shocked by these "revelations," although he was a convincing actor and could easily have been a silent partner in the conspiracy against Blomberg from the beginning. In any case Hitler lamented that "if a German field marshal will marry a whore, then anything can happen in the world."[22] That sealed Blomberg's fate.

Göring, acting on Hitler's instructions, told Blomberg that because of his disgraceful choice of spouse, he would have to resign from his position as war minister and from the army. Blomberg said "it was the greatest astonishment of my life and the worst blow I had ever received."[23] The savaging of Blomberg's reputation and the destruction of his career were generally greeted by silence or quiet assent from most military officers. Blomberg's lack of personal stature and political independence had won him little admiration from his colleagues.

With Blomberg gone, Göring's path to the War Ministry was blocked only by Colonel General Fritsch. Unless he found some way to dishonor Fritsch, that general would naturally succeed to the vacant office. As an army commander with a spotless reputation who was held in high esteem by his colleagues, Fritsch's elevation seemed to be a foregone conclusion. Göring turned to his allies, Heinrich Himmler and Reinhard Heydrich, respectively the commander and deputy commander of Germany's feared SS and Gestapo.

Oster had found out about the breaking scandal from his friend Gisevius. "I have just seen the fingerprints of your field marshal's wife," the former police official told him in reference to the unsavory rumors circulating about the young woman. "What, is that issue around again?" Oster replied. Then both went to see Canaris. Before they could say a word, Canaris burst out, "Isn't this all dreadful?" In the course of their conversation Canaris said ominously, "There's supposed to be something wrong with Fritsch, too."[24]

JANUARY 26, 1938, REICH CHANCELLERY

The scene was bizarre, even by the standards of the Third Reich. Colonel General Werner von Fritsch, commander in chief of the German army and recent critic of the führer's expansionist dreams, was summoned to the Reich Chancellery by Adolf Hitler. Colonel Hossbach, Hitler's adjutant, had warned the general the day before that he was about to be confronted by a man who had accused him of committing an unspeakable act. Hossbach now met the old soldier, who muttered gruffly, "This swine I must see by all means," and proceeded to the library with Hossbach.[25] Having served as the führer's army commander for the past four years, Fritsch believed that he was ready for anything.

Werner von Fritsch was a model German army officer. He had compiled a superb record as a student at the *Kriegsakademie* (War College), and had distinguished himself as both a staff and line officer in World War I, in which he had suffered a grenade wound. Like many of his brother officers, Fritsch occasionally expressed his distaste for the Nazis, although he generally approved of Hitler's foreign policy. Fritsch had overseen the rapid buildup of the Wehrmacht and had coexisted peacefully with Hitler until 1937, when he became concerned about the growing power of the SS and about Hitler's "rash" foreign policy moves.

In the library Fritsch found both Hitler and his chief lieutenant, Hermann Göring. What Fritsch could not know was that he was the target of a carefully scripted play that was about to unfold. Hitler led the two generals to a landing at the top of a staircase. Then, at a signal from Hitler, two Gestapo agents escorted onto the staircase Otto Schmidt, a denizen of Berlin's criminal underground. Schmidt had had a varied career as a blackmailer, police informer, extortionist, perjurer, and male prostitute. In 1936, when police were squeezing Schmidt for information, he mentioned that three years earlier, he had witnessed a high-ranking German officer engaged in a homosexual act behind the railway station in Potsdamer Platz with one "Bayern Seppl" ("Bavarian Joe"), a notorious male prostitute. Schmidt had followed the two to their outdoor assignation, a hurried tryst in a dark passage, and then proceeded to blackmail the officer for the next several months. Schmidt said that the officer's name was Fritsch. Although Hitler had been told about the allegation in 1936, he had brushed it aside and ordered Himmler and Heydrich to destroy the file. However, in January 1938 a changed political situation suddenly made the information useful.

As Schmidt approached the group on the landing, Hitler asked, "Is this the man?" Schmidt replied solemnly, "Yes, it was him." The general was so taken aback by this exchange that he was struck dumb. When he recovered his powers of speech he gave Hitler his word of honor that he had never seen the man. Hitler simply brushed this aside, as though the word of honor of his leading general counted for nothing against the accusation of a disreputable creature like Schmidt. Göring, meanwhile, played the part of a Greek chorus. He ran from the confrontation, loudly lamenting to all who could hear him, "It is he! It is he! It is he!"

Hitler told Fritsch that he now must resign from his post and from the army. If he did so quietly, the führer promised, there would be no disgrace. However, Fritsch recovered his wits sufficiently to reject this suggestion. "I refuse to account for myself in this way," he said. "These are criminals. I demand a court of honor." While Hitler desperately wished to avoid this alternative, he could not do so if Fritsch continued to insist on it. For fear of alienating the officer corps, Hitler could not risk such an affront to the honor of the army's commander. Finally the confrontation ended with Fritsch, still thinking that it was possible to rectify this hideous mistake, agreeing to a humiliating interrogation at SS Headquarters on Prinz Albrecht Strasse.

Fritsch labored under the delusion that the allegation against him, a case of mistaken identity, would be clarified once he had the opportunity to confront his accuser. Even Hitler expressed surprise at Fritsch's passivity during the entire encounter. He expected that Fritsch, as a proud German officer, would throw down his sword at Hitler's feet and stalk out of the room. To a man so given to tantrums himself, Fritsch's dignified response seemed incomprehensible.

The apolitical general was slow to understand the forces that were arrayed against him, and the motivation of his enemies. He correctly suspected that Göring and the SS were behind the plot, but he could not imagine that Hitler was party to it. After all, Fritsch reasoned, if the führer had wanted him to resign, he had only to ask. Throughout his entire ordeal, Fritsch never understood that the real enemy was Hitler himself, and the ultimate goal was not his disgrace but Hitler's seizure of command of the Wehrmacht.

January 27, Berlin, Army Headquarters

The day after the confrontation at the Reich Chancellery, Colonel General Ludwig Beck, chief of the General Staff, met with Canaris and Oster. Both Abwehr officers shared Beck's concerns about the attack on Fritsch. Their partnership had begun in 1935 when Canaris had been given command of the Abwehr. At that time Oster was a civilian employee of the Abwehr who had been stripped of his military rank when he had had an affair with the wife of a superior officer. Canaris, who had met Oster in 1931, quickly restored him to military rank and allowed him to build a network of anti-Nazi resisters within the Abwehr. One of these recruits, Franz Maria Liedig, noted that "Oster's office was a port of call for all those members or associates who were self-acknowledged opponents of National Socialism."[26]

Although he tolerated Oster's machinations, Canaris had a more ambivalent attitude than did his aide about Hitler and the Nazis.

Canaris was a resolute anti-Communist, and initially he welcomed the Nazis as nationalists who would protect Germany from a Communist takeover and restore the military to its rightful position of authority within the state. However, the guerrilla wars that Canaris fought against his professional rivals in Himmler's SS and Gestapo had dimmed his enthusiasm for the Nazi Party. And Hitler's increasingly bellicose foreign policy moves made Canaris apprehensive. But Canaris continued to support Hitler, at least until the Fritsch crisis opened his eyes. His close friend Lieutenant Commander Richard Protze said of the Fritsch crisis, "If you're looking for one specific event that shook Canaris' allegiance to Hitler, there you have it."[27]

Both Oster and Canaris agreed that the previous day's confrontation was really an attempt to sully the reputation of the commander in chief as a prelude to an attack against the army itself. Beck, who shared this view, argued forcefully that Fritsch had made a crucial mistake in allowing himself to be questioned by the Gestapo; he said that Fritsch should have refused to participate in the interrogation and should have gone to his military colleagues for help. Furthermore Oster and Canaris knew that, had Fritsch come to the Abwehr, they could have provided immediate and strong support.

Oster immediately saw that the Nazis' clumsy attempt to frame Fritsch provided an opportunity for action. Oster had served under Fritsch in the 1920s, and the two men had enjoyed a close personal relationship ever since. When he heard of the scandalous attack on his former commander, Oster said quietly, "I have made the Fritsch case my own."[28]

By 1938 Oster and his Abwehr colleagues had forged alliances with several police officials. The most important was Arthur Nebe, head of the Kripo (Criminal Police). Because of his high position, and the relatively free exchange of hallway gossip between Gestapo and Kripo officials, Nebe found out the truth and told Oster and Canaris.

The Gestapo knew that Otto Schmidt's allegations rested on the homosexual act that had been performed by "Bavarian Joe," the street name of Joseph Weingarten. When the hapless Weingarten was run to ground by the Gestapo, he was unable to identify the photograph of General von Fritsch as one of his clients. Furthermore, Schmidt's story did not check out. He alleged that he had followed "Fritsch" to his home in the Lichterfelde district of Berlin, where, in response to his blackmail threats, the officer had gone to a nearby bank and made several withdrawals from his account. An investigation by Gestapo agents revealed that General Fritsch had never lived in that part of Berlin, and did not have an account in a nearby bank. However, the investigation revealed that a retired army officer, Captain Achim Frisch, lived in the neighborhood and fit the profile. On January 15, Gestapo agents interviewed the ailing Captain Frisch, who acknowledged that he had committed the act with Weingarten and had given in to Schmidt's blackmail. Thus twelve days before Hitler's confrontation with Fritsch, the Gestapo knew that the charges against him were false, and Nebe passed this information on to Oster, whose suspicions were now confirmed.

Fritsch, however, was paralyzed by the turn of events. The January 26 confrontation in the Reich Chancellery and Hitler's repudiation of him had shaken the general and plunged him into a deep, nearly suicidal depression. Oster and others urged Fritsch to convoke a meeting of the *Generalität* (corps of generals), explain what had happened, and give them his word that the charges against him were false. If Hitler persisted in trying to remove him from his command, Oster urged him to call out the army and to resist by force. However, although Fritsch was a highly respected military man, he was incapable of fighting the intense political warfare that would be necessary to defeat his enemies.

JANUARY 28–FEBRUARY 2, BERLIN AND THE PROVINCES

Oster and Canaris understood that Hitler was about to decapitate the army, as he had beheaded the SA in 1934, but what could they do? Without troops under their command they were powerless. General Witzleben, Oster's friend and the commander of the Berlin military district (Wehrkreis), was seriously ill and hospitalized. The only recourse that Oster saw was to send out "missionaries" to enlighten provincial commanders. Germany was divided into twelve military districts. In a potential showdown, these district commanders would be the most important military men in Germany for, unlike Abwehr or General Staff officers, they had troops under their direct command. If a coup was to succeed, it would be because these generals supported it.

By January 28 Oster and his civilian collaborators, Hans-Bernd Gisevius, Carl Goerdeler, and Hjalmar Schacht, decided to alert friendly army commanders to the situation. Schacht, the highly respected former minister of economics and current president of the Reichsbank, was also energized into opposition over the Fritsch affair. "We all realized now that the leadership of the German Reich had fallen into the hands of common criminals." But out of "this terrible realization," Schacht claimed, "rose the still more terrible spectre of war."[29] He took it upon himself to seek interviews with Admiral Raeder and Colonel General Gerd von Rundstedt, the Wehrmacht's senior serving officer. But by the time he was able to speak with them, Hitler had gotten to the officers first, and had thrown sand into their eyes by claiming that he had other evidence of Fritsch's homosexuality, so that Schacht was unable to make progress. Indeed, when Schacht finally visited him on February 1 or 2, Rundstedt greeted him coldly, declaring "Mr. Minister, my lips are sealed."[30] An earlier meeting with Raeder had been similarly frustrating.

Oster, Gisevius, and Goerdeler received warmer receptions in the provinces. On January 29 all three met with local commanders of some of Germany's military districts. Oster journeyed to Hannover to meet with General of Infantry Alexander Ulex, commander of the XI Army Corps, and his intelligence chief, Major Alexander von Pfuhlstein. Oster first met with Pfuhlstein, who was sympathetic to Oster's position, and both men went to see Ulex. Oster briefed them about recent developments in Berlin, which were revelations to Pfuhlstein and Ulex. He suggested to the latter that he lead a delegation of all twelve district commanders to Hitler, demanding justice for Fritsch at the risk of their collective resignation. Ulex, though deeply shaken by the information, was hesitant to embark on the proposed mission, knowing that several of his fellow corps commanders did not share his hostility to the Nazis. Specifically Ulex believed that Generals Reichenau and Dollmann, commanders respectively of VII Corps (Munich) and IX Corps (Kassel) would not join an uprising against Hitler. But he brooded about it, even absenting himself from his command for several days until Pfuhlstein sought him out and begged him to return. He told the major remorsefully that "it is a great burden to me to have the feeling of having failed at a decisive moment."

Gisevius, Oster's friend from 1934 and a former Ministry of Justice employee, traveled to Münster, where he met with General of Artillery Günther von Kluge, the local Wehrkreis commander, and with Baron Hermann von Lüninck, provincial president in Münster and one of the few remaining anti-Nazis still active in Germany's civil administration. As Oster had found in Hannover, the two were completely ignorant about events in Berlin. Instead of being able to discuss concrete steps with them, Gisevius felt that he could do no more than present his case and allow it to sink in.

Meanwhile, Goerdeler, the former mayor of Leipzig, whose growing alienation from the regime the year before had pushed him out of office and into the arms of the resistance, returned to that city and contacted General of Infantry Wilhelm List, the local Wehrkreis commander; Major General Friedrich Olbricht, his chief of staff; and Lieutenant Colonel Edgar Röhricht, a new member of Olbricht's staff. They were in the middle of a staff exercise, and it was with some difficulty that Goerdeler detached them and filled them in on recent events. Goerdeler proposed that List march his troops on Berlin to "smoke out" the Gestapo headquarters, believing that if one brick was dislodged, the entire Nazi edifice would begin to crumble. List raised some logistical issues, pointing out that it would take three days to move an army division to the capital by train, and he could not do even that without assistance from his Berlin-based commanders. Furthermore the army would have difficulty defending itself against the Luftwaffe, which was certain to be hostile to an armed coup.

Nevertheless List was intrigued enough to travel immediately to Berlin, in company with Lieutenant Colonel Röhricht, to get an interview with Beck. He found Beck sympathetic but not inclined to revolution. List concluded that without leadership from their Berlin command, and most particularly the army's commander in chief, provincial generals could not possibly initiate an anti-Nazi revolution, no matter how strongly they felt about it. List had indeed learned the great lesson of the Fritsch crisis: Leadership for a coup must come from the center and the highest level of the army. If a coup was to succeed, it would have to have the full support of the army high command and be led by a troop commander in the Berlin district.

February 3, Berlin, Reich Chancellery

Fritsch still believed that Hitler was the unknowing dupe of the SS. Despite Oster's advice to the contrary, the general submitted to additional Gestapo interrogations because he believed that Hitler would see that he had nothing to hide and would exonerate him. Because Fritsch saw his dilemma primarily as a personal rather than a political issue, he was reluctant to engage the army on his behalf. It would be unconscionable, he told a friend, to begin a civil war over a personal issue. Thus on February 3, before Oster could mobilize army support behind him, the general gave in to Hitler's request and resigned, although he insisted on a court-martial to clear the slur on his name. Once he was no longer commander in chief of the army, Fritsch was powerless. Hitler could decide the succession at his whim and on his timetable.

February 4, Berlin

The day after he had bullied Fritsch into resigning as commander in chief of the army, Hitler stood before his *Generalität* and announced that he had abolished the position of minister of war. The führer himself would now fill that role. As a sop to the disappointed Göring, Hitler promoted him to the new rank of Reich marshal. The new commander in chief of the army was to be Colonel General Walther von Brauchitsch, a respected officer who had his own private problems. The general had been carrying on a long-term extramarital affair with a light-headed but charming widow, Charlotte Rüffer. Despite Brauchitsch's pleas for a divorce, his wife refused to agree unless the general gave her a capital sum that was far beyond his means. The despondent general contemplated resigning from the army. Knowing his vulnerability, Hitler entered into negotiations with Brauchitsch, promising him the money required for his divorce if the general agreed to conditions that virtually emasculated the army. The package of professional elevation and solution to his personal problems was too attractive for Brauchitsch to resist. He acceded to the führer's requests, and Hitler paid the general's wife a sum of eighty thousand Reichsmark, which allowed Brauchitsch to marry his inamorata.[31]

Colonel General Brauchitsch, who often seemed intimidated by Hitler, now owed his personal happiness to the führer. Brauchitsch was an old-fashioned soldier who, according to his adjutant, Curt Siewert, "saw the army only as an instrument of policy, but not a partner or rival on the same level [with the head of state]."[32] Although there were tensions between the new commander in chief and Hitler from the moment he took office, Brauchitsch shied away from open opposition to the führer's policies.

Hitler announced Brauchitsch's promotion and the accompanying changes in command structure before most general officers were aware of the events of the previous few weeks.[33] Hitler also used the confusion surrounding the Fritsch and Blomberg affairs to inflict still another massive and sudden change on the army—ordering the summary retirement of fourteen senior generals and the abrupt reassignment of forty others to different commands. There was no doubt in anyone's mind as to who was now in charge of Germany's armed forces.

Hitler's seizure of control of the Wehrmacht and the humiliation of the army leadership was complete. The two generals who had dissented at Hitler's November 5 meeting, Fritsch and Blomberg, were dismissed. The third dissenter, Konstantin von Neurath, was removed as foreign minister and replaced by the dull but obedient Joachim von Ribbentrop. Hitler had made himself head of the Wehrmacht, and the new commander in chief of the army was beholden to him. The führer had acted with a speed and decisiveness that would shortly characterize the Wehrmacht's new form of warfare: blitzkrieg.

Oster was despondent. He could see that Hitler had seized control of the army through a series of maneuvers based on deception and lies. Yet he was unable to organize the military strength to counter it. His missionary journeys seemed to have come to nothing, and he would have to fan the flames of conspiracy in another way.

March 17, Berlin, Preussenhaus

The end of the Fritsch crisis was as bizarre as its beginning. The court-martial that Fritsch had insisted on had been interrupted. During February and early March, Fritsch's legal defenders uncovered the entire seedy plot and the Gestapo's behind-the-scenes manipulations. The general's court-martial had begun on March 10, in the so-called Preussenhaus, which had once been the home of the Prussian House of Peers. The defense was certain that it would not only secure an acquittal for Fritsch but shake the foundations of the Nazi regime to its core.

But in politics timing is everything. The court-martial was adjourned the same day that it was convened when all military personnel were suddenly ordered to return to their units. Hitler had manufactured a crisis in foreign relations. By March 12 the Wehrmacht stood poised on the borders of Austria, prepared to invade if Austrian prime minister Kurt von Schuschnigg did not accede to Hitler's ultimatum to surrender control of his country to Austrian Nazis. The Fritsch crisis had been preempted by imminent war, which even his defenders recognized as the more pressing issue.[34]

By the time the court-martial resumed on March 17, Anschluss (union) with Austria had been peacefully achieved in what was called the *Blumenkrieg* (flower war). The führer had returned triumphantly to the country of his birth, where he had been received with enthusiasm bordering on hysteria. His prestige within Germany and his power within Europe had never been greater. The court-martial revealed all the deception behind the spurious charges against Fritsch and ended in his acquittal. The evidence clearly pointed to Himmler and Heydrich as the culprits. Göring feigned ignorance and expressed outrage at the fraud that had been perpetrated on the good general.

Urged on by Oster, Fritsch challenged Himmler to a duel with pistols. According to Otto John, a legal adviser to Lufthansa with ties to the resistance, "Oster . . . thought that this would bring about the decisive confrontation between Army and the SS and lead to the fall of the regime."[35] But Fritsch made the mistake of entrusting the delivery of his challenge to Rundstedt, who after the war admitted that he had simply carried the paper around in his pocket for a long time until he was able to persuade Fritsch to drop the matter.[36]

In urging Fritsch to fight a duel, Oster was clearly grasping at straws. The entire tawdry affair, which had revealed the duplicity of Himmler and the SS, had not led to the confrontation that Oster had hoped for. While some generals had become politicized by the Fritsch affair, others had lost their commands. Once more Hitler had seized the moment to thwart his enemies and reassert control over the army.

MID-MARCH, LONDON, FOREIGN OFFICE

For the second time in a year, civil servants at Britain's Foreign Office met with Dr. Carl Goerdeler, one of Oster's "missionaries." From 1937 onward, Goerdeler traveled regularly to France, the United States, and Britain, bringing the same message: There are Germans who oppose Hitler. Help them. Goerdeler always believed that reason could defeat the Third Reich. He was a man of energy, enthusiasm, and optimism whose faith, in retrospect, seems both endearing and sadly naive.

Goerdeler had first visited London in June 1937, bringing the news that Germany was seething with discontent and imminent civil war. He was received by members of the British Foreign Office like an alien from another planet. They were fascinated by his message but puzzled about the messenger. Goerdeler had characterized Hitler as "off his rocker," and predicted that "the Nazi dictatorship could not last more than six to eight months." He claimed that the army was ready to move against Himmler and the SS but was reluctant to take action against Hitler. However, he predicted that the army would eventually see that "to remove them [the SS leaders] without removing Hitler would be useless" because Himmler and Heydrich represented Hitler's deepest desires. He claimed that all outstanding issues between Germany and Great Britain could be resolved amicably "when the right men were in control in Germany."[37]

Goerdeler repeatedly promised Britain and France that if they stood firm against Nazi aggression, this act would support the German army and other moderate elements in their coming struggle against the Nazis for control of Germany. The Foreign Office personnel, in intraoffice notes, expressed their perplexity over Goerdeler. Orme Sargent, a senior civil servant, opined that "it is beginning to look as though Dr. Goerdeler had been sent to this country as the unofficial emissary and representative of the Reichswehr in order to put before H.M. Government the fears and wishes of the Army as contrasted to the Party." One of his colleagues noted warily of their loquacious informant, "I don't know much about Herr Goerdeler, but I am inclined to be rather chary of uninhibited advice from Germans."[38]

The first visit by Goerdeler had predated the Fritsch affair by six months. He came to London for a second time in mid–March 1938, probably at the instigation of Oster, Canaris, and Beck, *after* the Fritsch affair.[39] Many of the conflicts within Germany that he had mentioned earlier had begun to surface, and the international situation had become much more grave. Goerdeler made the startling claim that, if it had not been for Hitler's manipulation of the timing of the Anschluss, "an internal revolt would have been started in Germany by the Army; and Hitler and the Nazi Government would have been swept away." Goerdeler claimed that army discontent resulted from the revelation of SS fraud in the Fritsch case. While this was an overstatement, it contained a kernel of truth that his audience heard with curiosity and skepticism. As they had the previous year, senior British Foreign Office civil servants were dubious about Goerdeler. Ivo Mallet, for example, noted that he felt that Goerdeler, who "tends to exaggerate," was "over-optimistic" about the possibility of an army coup.[40]

These were certainly reasonable reservations to entertain about Goerdeler. Nevertheless the Foreign Office staff continued to listen to Goerdeler and to discuss his reports. By this means Lord Halifax, who was now foreign secretary, became aware of rumors of an army coup against Hitler.

April 24, Karlsbad, Czechoslovakia

Historical circumstances often take obscure men and thrust them onto the world's stage, however briefly. Such was the fate of Konrad Henlein, a former gymnastics instructor who on this day made the most important speech of his life. Henlein was addressing a rally of the Sudeten German Party (SDP: Sudetendeutsche Partei) in Karlsbad (today Karlovy Vary), a town in the far western part of Czechoslovakia that was overwhelmingly ethnic German.

Ever since the Anschluss, Henlein and the Sudeten Germans had agitated noisily for their own union with Germany. Since 1935, when the SDP had made a strong electoral showing under the leadership of the canny Henlein, Hitler had sent the party an annual subsidy of 180,000 marks. Henlein had caused as much trouble for the Czechs as he could, but opportunities had been limited until Germany's takeover of Austria in March 1938 provided him with the chance of a lifetime. Henlein, who swore to the British (and largely convinced them) that he was independent of the Third Reich, was in fact under Hitler's direct orders.

Hitler had summoned Henlein to Berlin on March 28, anointing him as the future governor of a Sudetenland repatriated to the fatherland, and had given Henlein his marching orders: "Demands should be made by the Sudeten German Party which are unacceptable to the Czech Government." Henlein, a backwoods rabble-rouser eager to demonstrate his grasp of the principle his führer had laid down, paraphrased what he had heard: We must "always demand so much that we can never be satisfied."[41] Hitler rarely smiled contentedly except in the presence of pretty women, dogs, or young children, but this certainly was an occasion when he might have. Henlein understood his role perfectly.

With his instructions from Berlin, Henlein's first attempt at formulating such demands came in a speech he delivered to his raucous followers on April 24. In his "Karlsbad program" Henlein laid down eight demands on the Czech government, at least two of which were calculated to be unacceptable. Sudeten Germans should have legal autonomy within the state, Henlein shouted, and "complete freedom to profess adherence to the German element and ideology."[42] He knew that these were two poison pills that the Czechs could not swallow, for they were an invitation to all the ethnic minorities in Czechoslovakia to press their claims for autonomy, which would inevitably lead to the disintegration of the country. No sovereign state could agree to them.

May 21–22, Czechoslovakia

With Austria swallowed whole by the Third Reich, Hitler eyed Czechoslovakia the way a hungry wolf eyes a crippled rabbit. Carefully scripted riots erupted regularly throughout the Sudetenland, ostensibly in protest against Czech oppression. Hitler publicly vowed to protect his "defenseless" fellow Germans.

The Sudeten agitation during the spring of 1938 reminded many observers of comparable events by Austrian Nazis on the eve of the Anschluss. Thus, on May 19, when Czech intelligence and French and British diplomats noticed increased German troop concentrations near the Czech border, they assumed that an invasion was imminent. Joachim von Ribbentrop, the German foreign minister, issued heated denials that this was true, just as he had on the eve of the Austrian invasion. Convinced that they faced certain war with Germany, the Czechs mobilized their forces on May 21. The next day both French and British diplomats delivered messages to Hitler. Halifax's note warned the Germans "not to count upon this country being able to stand aside if . . . there should start European conflagration." He added that "once war should start in Central Europe it was quite impossible to say where it might not end, and who might not become involved."[43] This indirect, even convoluted phrasing was vintage Halifax-speak. One can only imagine how it was translated into German. The warning message of the French was more direct about their intention to support their Czech allies.

The Germans did not invade. Most observers concluded that Hitler had been held off by rapid Czech mobilization supported by strong diplomatic warnings from the British and the French. In fact the May crisis was a misunderstanding, and for once Ribbentrop had been telling the truth. Hitler *did* intend to invade Czechoslovakia, but not in May. The reported troop concentrations had actually been Wehrmacht formations on routine spring maneuvers.

While the May crisis had been a false alarm, its consequences were grave and far reaching. Chamberlain and Halifax were frightened by the successful outcome of their diplomacy. The prime minister confided to his sister Ida that "the more I hear about last weekend the more I feel what a 'damned close-run thing' it was."[44] Chamberlain's brief flirtation with warnings to Germany was over.

Even Halifax was chastened by his own boldness, and startled by how it seemed to have pushed Britain to the edge of an unwanted conflict. He remarked to Sir Alexander Cadogan, his chief civil servant at the Foreign Office, that "we *must* not go to war!" and he made the same comment the next day to a cabinet member. On May 22, the day after Halifax had sent his warning to Hitler, he telegraphed the French that they could not count on British support in the event of war with Germany.[45]

In Berlin, Oster and the conspirators were heartened by the public British reaction to the May crisis. Of course they had no way of knowing how quickly Halifax had stepped away from his actions. They only saw a man who had apparently stood firm against Hitler, a course of action that they had urged on their British contacts over the last year. If he was to transform the growing discontent with Hitler's reckless policies into a successful conspiracy, Oster knew that the British would have to be persuaded to stand fast against Hitler's aggression.

MAY 28, BERLIN, REICH CHANCELLERY

Hitler had been infuriated by the May crisis. "The world Press jubilantly announced that the German dictator had yielded," remembered Paul Otto Schmidt, Hitler's translator. "Anyone deliberately planning to madden Hitler could not have thought of a better method."[46] He had no intention of invading Czechoslovakia at that time, but it appeared to all the world that he had been forced to back down. He remained in a funk for the rest of the month, dreaming of the revenge that he would wreak on the world with his panzers and bombers.

On May 28 he met with his military commanders. Standing before a large map of central Europe, he swept his hand across it in one all-enveloping gesture and declared, "It is my unshakable will that Czechoslovakia shall be wiped off the map."[47] Hitler predicted that France and Britain had no stomach for war. As long as the invasion of Czechoslovakia was accomplished quickly, they would accept it.

After his speech, Hitler approached Generals Beck, Brauchitsch, and Keitel, who were standing in a corner of the solarium. " 'So, first we will clear up the situation in the East, and then I will give you up to four years for preparations, and then we will deal with the West.' The generals listened in silence."[48] Undeterred by his top commanders' obvious lack of enthusiasm, Hitler ordered staff planning for war to be accelerated, culminating in an invasion of Czechoslovakia no later than October 1—"X-Day." The bomb was armed; the timer was set. The Wehrmacht would explode on Czechoslovakia, and perhaps the world, in four months' time.

Throughout the spring Oster had struggled against Hitler's attempts to disgrace his colleagues and seize control of the army. At every turn he had been foiled: His belief that the Fritsch crisis would provide a perfect chance to launch a putsch against Hitler had backfired. General Witzleben, the logical man to lead the putsch, had been incapacitated by a serious illness, and Oster's attempts to persuade provincial commanders to take the initiative had been unsuccessful. Hitler had outmaneuvered him on every front.

Nevertheless the May crisis created a moment of opportunity for Oster and the conspirators. "Hitler is vulnerable on the issue of Czechoslovakia," Oster said.[49] He thought that the apprehension of his military colleagues might be converted into determination to take action against the regime. Oster believed that in his rage against the Czechs and his contemptuous disregard of the British and French, Hitler had set the stage for self-destruction. A close friend of Oster's, Fabian von Schlabrendorff, described in his postwar memoirs the conspirators' thinking in the summer of 1938: "A tough stand against Hitler by the Western Powers would have strengthened our position immeasurably, and would have brought many still undecided or wavering generals into our camp."[50]

In Hitler's declaration of his intention to go to war on May 28, the führer had set the bomb ticking. Oster and his fellow conspirators knew that they had only four months to stop this madman from dragging them into a war. They would try to dissuade him or, failing that, to remove him. If it came to a coup, they knew from the Fritsch affair that they could succeed only with the support of the Army High Command. Foremost in their minds was a crucial question: Could they count on the support of Generals Beck and Brauchitsch?

2

The Road to Mutiny and Revolution:
General Beck Converts

Planning for war was the meat and potatoes of General Staff work, and none did it better than the General Staff of the German Army. With a glorious heritage that dated back to Gneisenau and Moltke, distinguished chiefs of the Prussian General Staff, members of the General Staff were rightly proud of the sage advice they had given to their political masters. The officers of the General Staff had never considered themselves mere functionaries, relegated to carrying out the orders of their rulers. They had always been consulted, and their advice carefully weighed, in any diplomatic decision that might result in war.

General Ludwig Beck was the embodiment of the revered office of Chief of the General Staff. The position had been Field Marshal von Moltke's during the wars of 1866 and 1870–71, and held enormous prestige. Beck was a brilliant, widely read man, especially in international relations, and a highly decorated soldier. Like Oster, Beck was an avid and accomplished equestrian, and the two often rode together. Beck's major accomplishment during World War I had been to plan the orderly withdrawal of the ninety divisions of the German army at war's end. In the 1930s he had authored the respected army manual of tactics, *Truppenführung*. Beck was already well known and well liked within the officer corps, but this elegant publication made him famous. Although he hated small talk and social events, he had great personal charm. This, along with his work ethic, his dignity, and his brilliance made Beck one of the most respected officers in the Wehrmacht. Oster knew that it was essential to win him over to the conspiracy.

Beck, like Fritsch, found the Nazis vulgar and distasteful. He was also deeply religious and politically conservative. In his approach to military matters he believed in proceeding with caution. In the wake of Hitler's meeting with his commanders on November 5, 1937, when Beck had learned of the führer's plan for the possible invasions of Austria and Czechoslovakia, he had been profoundly disturbed. In a memorandum written on November 12, 1937, Beck had attacked the basic assumptions of Hitler's plan, beginning with the presumed need for "Lebensraum," arguing that Germany's economic goals could just as easily be met by foreign trade.

In the same memorandum, Beck had also reiterated Blomberg's and Fritsch's warnings not to underestimate Britain and France: "The extent of French and English opposition to increases in German power and space should not be misjudged." Beck dismissed as "a most improbable daydream" Hitler's casual assertion of French military paralysis resulting from political turmoil. He pointed out that the French army was so strong that even in the case of a war with Italy, France would "always have enough forces to set against Germany."

Beck agreed that Austria and Czechoslovakia came "naturally" within Germany's sphere of influence and might eventually be incorporated into the Reich. But if Hitler acted recklessly, he would surely bring Britain, France, and perhaps even the Soviet Union into a war against Germany. And that would be a war, Beck strongly implied, that Germany could not win.[1]

Despite his reasoned dissent, Beck was not the sort of officer precipitously to desert his post and join a revolt against the regime. Indeed, when pushed by Oster and Beck's own deputy, General Franz Halder, to take a more confrontational role during the Fritsch affair, Beck had turned on them angrily and said, "Mutiny and revolution are words that have no place in a German officer's vocabulary."[2]

Meanwhile, in the early summer of 1938, Europe's political temperature remained high. While the May crisis had passed, the British feared that the Sudeten issue would arise again unless they intervened in some way. But what should they do? Chamberlain and Halifax cast about for a solution.

Hitler was determined to wage war no later than October 1. Oster knew that if he were to derail the führer's plans for conquest, it would be crucial to make Beck and then his superior Brauchitsch see that the only way to stop Hitler was to overthrow him. But the cautious Beck would not be won to the conspiracy overnight. If he was ever to reach Oster's conclusions, it would be as the result of his own intellectual journey.

JUNE 3, BERLIN, ARMY HEADQUARTERS, BENDLERSTRASSE

Beck's reaction to the May crisis and Hitler's announcement of his intentions was complicated. As chief of the General Staff, Beck saw his role as loyal adviser to the head of state. In this capacity it was his duty to warn the führer of the danger of his plan of action. He had done so in a memorandum of May 5, 1938, which reiterated the arguments he had made in his memorandum of November 12, 1937. When Hitler had read Beck's carefully reasoned argument, he had contemptuously swept it aside, saying that its data were "not objective."[3]

When he heard Hitler's announcement of war within the next four months, Beck returned to his desk. How could he persuade the führer of the suicidal consequences of his intended course of action? After working nearly continuously since the Reich Chancellery conference on May 28, Beck had prepared still another memorandum attacking Hitler's announced plan for an invasion of Czechoslovakia. Rather than focus on world political considerations, which had drawn Hitler's scornful dismissal, he now tried a different approach: The führer's invasion plans could not be implemented because they had been ordered without consulting the General Staff. "Without appropriate consultation," Beck asserted, "the Army General Staff cannot accept the responsibility for carrying out the orders."[4]

Beck was clearly grasping for a strategy that would convince Hitler to step back from his invasion plans, but he was limited by both his place in the military hierarchy, and his own rather rigid opinion of what constituted appropriate action. His superior, Colonel General Brauchitsch, shared his perception of the danger, but was easily intimidated by Hitler and gave Beck only lukewarm support. Furthermore Beck himself was far from ready to embrace the idea of a putsch. As much as he disliked Hitler, he still believed that the challenge was to persuade rather than overthrow him.

Meanwhile, Hitler was growing tired of this cautious chief of the General Staff. He told his aides that Beck was Fritsch's "evil spirit" (*böser Geist*) "because it was Beck who put the brakes on and Beck who presented memoranda that he considered lies." When one of his adjutants asked what he meant by Beck's lies, Hitler replied that Beck had exaggerated the size of the French army by including their reserves, while he had underestimated the size of the German forces by not including the SA and SS.

"In my eyes," the führer said, "that is a lie."[5]

June 3, London, Foreign Office

While Beck feared that Hitler's aggression would transform Britain into an enemy, members of His Majesty's Government struggled to find an appropriate response to Hitler's threats against Czechoslovakia. How could Britain support its ally France in its commitment to the Czechs while not provoking Germany into war? The struggle for a solution to this dilemma consumed British politicians throughout the summer and early fall of 1938.

Sir Robert Vansittart was the Foreign Office's chief expert on Germany and chief diplomatic adviser to His Majesty's Government, a grand title that masked a kick upstairs. Although "Van" recognized the demotion for what it was, he refused to retire gracefully into bureaucratic oblivion. As the Sudeten crisis deepened, he continued to pepper his masters with sharply worded memoranda.[6]

He wrote an annoyed note to Lord Halifax: "They are at it again in their leader today." The object of Vansittart's anger was the editorial board of the London *Times*. Taking a strong proappeasement stance, on June 3 *The Times* urged that the Sudeten Germans "ought to be allowed by plebiscite or otherwise to decide their own future, even if it should mean their secession from Czechoslovakia to the Reich." At the moment no one in His Majesty's Government was advocating cession of the Sudetenland to Germany. However, *The Times* was widely regarded as the unofficial outlet for government opinions. As Englishmen took their morning tea and *The Times*, they must have wondered whether this lead editorial was an indication that their government was planning to endorse this extreme measure.[7]

EARLY JUNE, BERLIN, SD HEADQUARTERS

Hitler's campaign against Czechoslovakia depended on a public perception that the Czechs were heartlessly oppressing their Sudeten German subjects, and that the only solution to the Sudetens' misery was intervention by Germany. To orchestrate their end of the show, the Sudeten leaders secretly traveled to Berlin throughout the summer to receive instructions from SS operatives.

Konrad Henlein was in Berlin to consult with Werner Lorenz, an SS general of police whose responsibilities, as head of the *Volksdeutsche Mittelstelle* (Center for Ethnic Germans), included ethnic Germans abroad. Henlein unburdened himself to Lorenz. "What attitude should I adopt," Henlein asked, "if the Czechs, under foreign pressure, suddenly accede to all my demands and present, as a counter-demand, entry into the Government?" Henlein stewed over this tactical dilemma for a moment, and answered his own question: "I shall answer 'Yes' with the demand that the foreign policy of Czechoslovakia [toward Germany] should be modified. The Czechs will never agree to that."[8]

JUNE 13, BARTH AIRFIELD, NEAR STRALSUND

Hitler assembled the same group of generals that he had addressed on February 4 about his decision to remove Fritsch and to replace him with Brauchitsch. The dishonor done to Fritsch, particularly after his court-martial, still rankled among his colleagues and supporters. Captain Gerhard Engel, Hitler's army adjutant, claimed that "at every opportunity, day and night, I fought for the rehabilitation of Fritsch."[9] Hitler must have felt that some ameliorative gesture was needed, especially with "Plan Green"—his plan to invade Czechoslovakia—in the wings.

According to General of Infantry Curt Liebmann, Hitler told his audience that he had come to the conclusion that the Czech question "could only be settled by force."[10] This came as a surprise to Liebmann, who before then "was not aware that there actually *was* an acute Czech question." Hitler also mentioned that he was in the process of strengthening the western fortifications, and had put Colonel General Wilhelm Adam in charge.

Then Hitler settled in to discuss the Fritsch case. "The judicial injustice against Fritsch ended with his exoneration," the führer said somberly. "This episode now looks like a true tragedy." Hitler pointedly refused to take responsibility. While he regretted what had happened to Fritsch, Hitler said, he "could not have acted differently." He promised that Fritsch would receive "every satisfaction," although reassignment to his old position was out of the question. After these few words, Hitler left the meeting.

The remainder of the meeting was conducted by Major General Walter von Heitz, head of the Reich Military Court, who explained how the accusation against Fritsch had been based on mistaken identity. Undoubtedly his listeners smelled an SS rat behind the attempted frame-up. Liebmann noted that while he was pleased to know the facts of the case, "the issue for me remained unresolved." Two days later Hitler "elevated" Fritsch to honorary command of his old unit, Artillery Regiment Twelve. One can only imagine how many of the other senior officers at the meeting shared Liebmann's disquiet at the shabby treatment meted out to their esteemed colleague.[11] In his obvious contempt for Fritsch's honor, and by implication for the honor of the army, Hitler was sowing the seeds of discontent among the *Generalität* that Oster hoped to reap.

MID-JUNE, BERLIN, ESPLANADE HOTEL

General Beck addressed his fellow General Staff officers at a dinner at the conclusion of an indoor war game, a traditional exercise in which officers explored the likely outcome of a hypothetical conflict. Having failed to move Hitler with his memorandums on the dangers of the planned invasion of Czechoslovakia, Beck struck from a different angle. The war game had presumed a Wehrmacht invasion of Czechoslovakia with the assumption that France would attack Germany from the west. Beck showed that the exercise had demonstrated that while the Wehrmacht could defeat the Czechs in three months' time, the French would advance deep into German territory, portending eventual German defeat. Beck drew the conclusion that the Reich's political leaders must be made to see the folly of their present expansionist plans.

Beck's intent was clear, but his speech received a mixed reception. Lieutenant Colonel Edgar Röhricht, a participant and Beck sympathizer, noted that many junior officers disagreed with the chief's conclusions. Rudolf Schmundt, once an admirer of Beck's, now Hitler's new adjutant, claimed that he "did not understand the dynamism of the new regime. . . . If it were up to Beck, we would still be begging at the conference tables of Geneva." A junior Luftwaffe officer guffawed. "More than thirty days to overrun these ridiculous Hussites! As if we had no air force!" Despite this dissent the mood of the dinner was somber, and more officers agreed with Beck than did not.[12]

The war game was one of Beck's several attempts to warn Hitler of the dangers of war over Czechoslovakia. The führer's reaction was to denounce these as cowardice. "What kind of generals are those that I have to drive to war! Is it right that I should have to drag the generals into war?" he fulminated. Hitler did not want analysis, only assent. "I do not require that my generals understand my orders; only that they obey them."[13]

At about the same time that Beck was conducting his war game in Berlin, Lieutenant Colonel H. C. T. Stronge, the British military attaché in Prague, was in London for the annual conference of British military attachés. Stronge had served with distinction as an infantry officer during World War I, and had made his postwar career largely as an intelligence officer. Given his posting and the developing Czech crisis, Stronge noted that the conferees were particularly attentive to what he had to say.[14]

Earlier that April, Colonel Stronge had been invited by the Czech High Command to tour fortifications on the German frontier. He was given three days to inspect any area he wished. Surprised by this unusual invitation, Stronge quickly took advantage of it. He had been greatly impressed by the defensive strength of the whole system, which had been skillfully constructed. He thought that "even at that comparatively early stage the works would constitute, if manned, an obstacle of some delaying power to an attacking force." By autumn, after another six months of construction, he thought that the fortifications would become "truly formidable."

In London, Stronge shared these views with his colleagues, adding also that the morale of the Czech army was very high, as was its general preparedness for war. At the close of the conference, Colonel Stronge was summoned by Leslie Hore-Belisha, secretary of state for war, and by Lord Halifax, for separate interviews. "At these several interviews I gave it as my opinion that the army would fight it out if allowed to do so by the Prague government," Stronge remembered. He added that if the Czechs were able to successfully resist the first German attacks, it "might well have a decisive effect upon its [the war's] further course."

What specifically interested both Hore-Belisha and Halifax was how long the Czech army could keep the Wehrmacht at bay. When pressed by Halifax about the minimum amount of time that the Czechs could hold out, "I hazarded an estimate of three months." Of course, Stronge added, if France and Britain, and perhaps the Soviet Union, joined the fray, the incident would escalate into a world war, and there was no predicting where it might end. But, he concluded, "the more I saw of the [Czech] troops in 1938, the General Staff and the whole of the military ensemble, the more convinced did I become that here was a force capable of defending its frontiers and willing to do so at any cost."

Thus Colonel Stronge and General Beck, independently examining the Wehrmacht's likelihood of success, arrived at almost exactly the same conclusions. The Czechs *by themselves* could hold off the Germans for three months.[15] And if the French and British came to the Czechs' assistance, the Wehrmacht would be facing significant problems. In Germany, Hitler disregarded Beck's analysis, while in England, Halifax disregarded Stronge's.[16]

This miscalculation of the probable success of a Wehrmacht assault against Czech defenses, shared by both Hitler and Halifax, was disastrous. It increased the possibility of a German attack and decreased the possibility of a vigorous British response.

JUNE 30, THE BERGHOF

Colonel General Wilhelm Adam, fourth in seniority among the generals, was the sort of soldier Oster would have to recruit to the conspiracy if it was to succeed. The old general marched stoically into Hitler's country house at Berchtesgaden. He knew that he faced a confrontation with the führer, who had been alarmed by a scathing report from Göring on military preparedness. At the end of May, when Hitler had decided that he would make war against the Czechs by October 1, he had reassessed the military situation. Since Germany and Austria now surrounded Czechoslovakia on three sides, the Wehrmacht could choose among many invasion options. Hitler felt confident about the attack. But what about Germany's western defenses? If the French fulfilled their treaty obligations to Czechoslovakia, the only way for them to strike was across the Franco-German border, which was virtually unfortified on the German side. The Germans would have to hold the line with a handful of divisions against the main strike force of the French army. The Wehrmacht's sword was strong but its shield was nonexistent. Hitler had neglected these fortifications—variously called the "West Wall" or the "Siegfried Line"—because he had not envisioned an early war. After the May crisis his calculations changed.

Typically, Göring had stuck his nose into this issue. In early June he had conducted an "inspection tour" of the western fortifications in his luxury train, accompanied by a huge entourage. The Fat One had sent a report to the führer that, according to General Keitel, "was one long accusation against the Army from beginning to end; virtually nothing had been done, what had been done was inadequate, and there was hardly the most primitive system."[17] Göring's motive, as usual, had been to embarrass his army colleagues. In any case he had succeeded in alarming the führer, who summoned Adam to the Berghof.

Adam was a sixty-two-year-old army officer who had served as chief of the General Staff in the last years of the Weimar Republic, and had been posted to both Munich and Berlin. Nazi leaders who knew him distrusted him because of his political independence, but he was highly regarded by his brother officers. In the event of war Adam would command all the German troops on the western front. In late April he had been given responsibility for constructing German fortifications in that sector. After the May crisis, Hitler had shifted fifty thousand troops to Adam's control, and told him to finish the fortifications by autumn. Adam recalled that "at first I took this task humorously, and told Brauchitsch that it was a matter of luck how many bunkers could be constructed by autumn."[18]

Adam was not the sort of general whom Hitler could bully. Their meeting was tense. The führer told Adam that he wanted ten thousand concrete bunkers and eighteen hundred heavy gun emplacements—to protect eleven divisions—constructed no later than October 1. Adam flatly replied that it could not be done. By October 1 the best that he could hope for was a comprehensive plan for fortifications, and the beginning of construction. Furthermore Adam disagreed with Hitler about the strategy behind the fortifications. Adam favored a "fortified area" plan that would build defenses behind the frontier in places most easily defensible, whereas Hitler imagined a straight-line fortified boundary that resembled the trenches in which he had fought as a soldier in World War I.

The conflict between General Adam and Hitler about military fortifications was, at a deeper level, a conflict over political strategy. Adam thought that war in the autumn of 1938 would be suicidal. He wanted to build a fortification system that would be useful in the distant future, when Germany might be ready to confront its enemies. Hitler could not wait. He was planning on war no later than October 1. He would take whatever could be built by then, partially as a military defense but mostly to deter the French from attacking him.[19]

Hitler ended the meeting by saying, "In any case the bunkers must be completed in three months." Adam ruminated, "This man could never be taught anything." The next morning over breakfast, having insisted that Adam stay at the Berghof overnight, Hitler ranted about the Western Powers. "I will get England yet," he concluded ominously. Adam left the Berghof pleased that he had spoken frankly to Hitler. He hoped that he had had some influence on the führer's limitless ambitions. Perhaps, he thought, Hitler will delay his "robbery plans" for a year.[20]

Adam was vocal to the point of indiscretion about his objections to Hitler's plan. Oster must have marked him out as a potential resister. It would not be long before the outspoken general would be approached about cooperating with the conspirators.

JULY 3, LONDON, APARTMENT OF T. PHILIP CONWELL-EVANS

Feeling that he needed reliable information about British intentions, Beck approached Erich Kordt and asked him to use his connections in England. This is the earliest indication of Kordt's being drawn into the military resistance, although it is likely that he had some previous indication of the generals' growing discontent.

Kordt traveled to London to meet with his old friend T. Philip Conwell-Evans. Beck was hoping to establish a connection to British policy makers through him. Although Erich Kordt was relatively junior in the diplomatic corps, he was influential because of his close connection to the limited and boorish Joachim von Ribbentrop, whom Hitler had recently elevated to the position of foreign minister. In Ribbentrop's meteoric rise in the Nazi hierarchy, he had come to depend on the clever young Kordt to repair his fractured prose and to tidy up his various messes. Through Ribbentrop, Kordt had access to the Reich Chancellery and insider information, and also had freedom of travel.

Erich Kordt and Conwell-Evans had become friendly through Erich's brother, Theo, who had attended university in Britain, and had served an earlier term in the German Embassy in London. Conwell-Evans—the most significant of Vansittart's "German spies"—had been a lecturer at Königsberg (now Kaliningrad) University and had maintained his ties in Germany. Kordt explained that "we decided to inform the British Government of Hitler's plans [for war against Czechoslovakia], and we would try to influence the British at the proper time which position to take in the event of a German-Czech crisis." Beck had moved to a more radical position than the one he had held in June, since he had instructed Kordt to convince the British that "a firm declaration by Britain would place him in a position where he [Beck] could incite the Army to active revolt against the regime should Hitler go to war against Czechoslovakia." Kordt urged Conwell-Evans to pass on this message to members of the government.[21]

July 16, Berlin, Army Headquarters, Bendlerstrasse

Beck had written still another memorandum and took it to Brauchitsch, who discussed it with him the same day. Beck reprised his earlier arguments. War against Czechoslovakia would inevitably bring in Britain and France, which would mean Germany's defeat in a prolonged world war. Since his previous memorandums had been either ignored or dismissed by Hitler, Beck had begun to think of what other measures he could take to stop the führer in his reckless march to war. In this memorandum Beck considered the responsibility of soldiers: "History will indict these commanders of blood guilt if, in the light of their professional and political knowledge, they do not obey the dictates of their conscience." Knowing that he was treading on a delicate issue, particularly in an army famous for its iron discipline, Beck concluded, "The soldier's duty to obey ends when his knowledge, his conscience, and his sense of responsibility forbid him to carry out a certain order."[22] Beck was pondering under what circumstances he could justifiably set aside the loyalty oath he had sworn to Hitler in 1934.

As Beck edged ever closer to revolt, Hans Oster whispered in his ear. General of Artillery Halder, then Beck's deputy, remembered that, during the summer of 1938, Oster "had gradually become the most trusted person for Beck on matters of resistance." He noted that Oster was "a frequent guest in Beck's office. The meetings between the two of them lasted for hours." As Beck became increasingly alarmed by the direction of Hitler's policies, he listened sympathetically as Oster urged him to embrace revolt.[23]

JULY 18, LONDON, HOME OF LORD HALIFAX, NUMBER 88 EATON SQUARE

As the weeks slipped by, Hitler, like Beck, wanted to know what the British were thinking. Two weeks after Beck had sent Erich Kordt on his secret visit to Conwell-Evans, Hitler dispatched his adjutant, Captain Fritz Wiedemann, on a clandestine mission to London.

With its stately homes and well-kept gardens, Eaton Square in the fashionable Belgravia neighborhood was perfectly situated for an ambitious young Tory politician. With the good fortune that seemed to bless him throughout his life, young Edward Wood, eventually to become Lord Halifax, had inherited the magnificent house at Number 88 Eaton Square from his aunt. It became his London residence, an easy commute to his workplace in 1938, the Foreign Office in London's Whitehall.

At 10 A.M. Halifax received Wiedemann at his home. The only other person present was Sir Alexander Cadogan, permanent undersecretary of the Foreign Office, who was there to translate. The meeting between Halifax and Wiedemann was meant to be secret, but a reporter from the tabloid *Daily Herald* had recognized the German when he landed at Heston Airport in Hounslow, West London, and wrote a story about it the next day.

Wiedemann settled in front of the fireplace at Number 88, communicating with Halifax and Cadogan "in a very open, trusting manner."[24] On Hitler's specific instructions, the meeting was being conducted without Ribbentrop's knowledge. When Halifax raised an eyebrow about these unusual arrangements, Wiedemann leaned closer to him and said in confidence, "Herr Ribbentrop's position with the Führer is no longer what it had been."[25]

During World War I, Wiedemann had been an officer in the regiment in which Hitler had been a mere message bearer. In answer to a question by Halifax, Wiedemann said that Hitler had been a "brave, reliable, and cool soldier," but he had never seen any indication of the corporal's future greatness. After the war Wiedemann had returned to farming, but he had joined Hitler's staff in 1934 and was now privy to secrets of the Nazi inner circle.

Hitler had sent Wiedemann to meet with Halifax ostensibly to discuss the possibility of a Göring visit to Britain sometime in the near future. However, both Halifax and Wiedemann were sensitive enough to recent tensions over Austria and Czechoslovakia to know that such a visit was highly improbable. Only Chamberlain continued to think that it was a live option but even he conceded that it was unlikely to take place before autumn.

While there was desultory conversation about the anti-Nazi tone of the British press and the Anglo-German naval treaty, the real business of Wiedemann's visit was to probe British intentions toward the Czechs.[26] Hitler had given Wiedemann detailed instructions, telling him to emphasize to Halifax that if the Sudeten issue was not settled peacefully, he would resort to force. Hitler had said that the Sudeten Germans wanted to return home to the Reich and were being stopped from doing so by the Czechs. "I am not going to wait much longer. If no satisfactory solution has been achieved soon, I will solve this question with violence. Tell Lord Halifax that!"[27]

According to Wiedemann, he meticulously repeated Hitler's message. Halifax, anxious to pin Wiedemann down, asked about specific German intentions toward Czechoslovakia. "Could we get a binding declaration from the German government that it will not solve the Sudeten-German question with violence? A solution like that [would] not be taken well by the English people." Halifax repeated the question slowly in several different forms; Wiedemann replied that he could not give such a blanket assurance. He added that Hitler certainly would not sit still and watch while the Czechs started riots against Sudeten Germans.

"How long do we have to solve the problem peacefully?" Halifax asked quietly. Wiedemann parroted Hitler's deliberately misleading timetable: "Approximately until March 1939."

"Until then we can achieve a lot," Halifax answered hopefully, and the topic was dropped.

In his report of the meeting, Halifax remembered parts of the conversation somewhat differently than did Captain Wiedemann. The foreign secretary claimed that Wiedemann had given him "binding assurances" that the German government was "planning no kind of forcible action." He also said that this assurance would be "limited to a definite period" but he made no mention of a specific date. Later in the two-hour conversation, Halifax claimed, Wiedemann had again repeated "most emphatically that the German government was planning no resort to force." However, the captain had tacked on a crucial qualifier. Germany might be forced to take action "by some unforeseen and serious incident," such as the "massacre" of Sudeten Germans. Halifax reported Hitler's message as less threatening and more promising than it actually was, for by that time he and Chamberlain were cooking up a scheme for the British government to intervene in the growing Czech crisis.

Chamberlain then imposed his own rosy spin on Wiedemann's statements, which further distorted the führer's message. During a cabinet meeting two days later, Chamberlain—who had obviously been briefed about the meeting by Halifax—failed to mention Wiedemann's proviso about "incidents" that might force Germany to abandon its pledge of "no forcible action." Chamberlain also claimed that "he gathered that the period contemplated might be one of a year," although Wiedemann had said nothing of the kind.[28] Chamberlain's interpretation was consonant with his own optimism about Anglo-German relations.

When Wiedemann returned to Germany, he also changed the message, telling his colleagues that Britain was determined to resist German aggression over Czechoslovakia, an exaggeration of Halifax's remarks that Wiedemann hoped would discourage Hitler's adventurism. But in the end the adjutant's foray into creative diplomacy was doomed to failure. Writing to Halifax after the war, Wiedemann said "after my return to Germany Hitler would not anymore listen to my report."[29]

Despite being frozen out by Hitler, Wiedemann rendered the conspirators a valuable service during the summer of 1938. He was friendly with Hans von Dohnanyi, one of Oster's collaborators in the Ministry of Justice, and through him, with Oster. Wiedemann remembered that "Oster and Dohnanyi visited me pretty regularly in order to inform themselves about Hitler's opinions." Wiedemann supplied this information freely and was trusted by the conspirators, although it is unclear precisely how much he knew of their plans.[30] Nevertheless, in Wiedemann, Oster had an informant at the center of Hitler's entourage.

July 19, Berlin, Army Headquarters

Beck finally decided to speak to Brauchitsch in a new, forceful way. According to a Beck biographer, "Oster must have impressed upon Beck again and again the need for more forceful measures. . . . In the end Beck gave in to Oster, albeit with reluctance."[31]

Beck's new plan for thwarting Hitler's intentions considered the possibility of a violent showdown between the army and the National Socialist regime. Beck's memorandum declared that in order to prevent a war against Czechoslovakia "all conceivable ways and means down to the ultimate consequence must be applied," and the question should be examined whether or not his proposed démarche against war "should be activated to the extent of letting it come to a conflict with the SS and the bigwigs." However, Beck was not yet ready to present Hitler as a villain, but rather as a pawn in the hands of evil forces. He emphasized that "there must be no doubt that this struggle is on behalf of the Führer." To set the stage for this confrontation, in which the army leadership might set itself against the most widely hated aspects of Nazi rule, Beck suggested the following platform: "Short, clear catchwords are in order: 'For the Führer, Against War, Against Bossism (*Bonzentum*), Good Relations with the Church, Freedom of Opinion, An End to Tcheka Methods, Justice. . . . No Palace-Building, Public Housing, Prussian Purity and Simplicity!' "[32]

Beck's sloganeering was an uncharacteristic public relations ploy, reflecting not his actual thoughts but a "line" he thought would play. In its awkward attempt to deflect attention away from Hitler, it indicates the difficulty that many resisters had in contemplating a direct attack on the führer. Partially this was because they were reticent themselves, and partially it was because they knew that while they might be able to persuade their colleagues to take on the SS, it would be an altogether different matter to convince them to attack Hitler directly. For many German officers, a willingness to revolt against the regime stopped at the body of the man to whom they had taken a loyalty oath.

For Beck, however, Hitler was clearly the enemy, and his expansionist plans had to be stopped. He wrote to Brauchitsch that "there must not arise the remotest suspicion of a conspiracy," although that is clearly what he intended. It was essential, he said, to maintain the "united stand [*Geschlossenheit*] of the highest military commanders" against Hitler's war plans.

Brauchitsch's reaction to Beck's tirade is unknown, but Oster must have been satisfied to see Beck edging ever closer to revolt.

JULY 21, BERLIN, FOREIGN MINISTRY

General Beck was not the only one trying to avert war. At the Foreign Ministry, Ernst von Weizsäcker was growing increasingly concerned about

Hitler's foreign policy. Weizsäcker was a retired naval officer from a distinguished family who had been elevated to the position of *Staatssekretär* (state secretary)—equivalent to the British permanent undersecretary—in the Foreign Ministry, making him Ribbentrop's chief deputy.

Ribbentrop called in Weizsäcker to talk about the role of the Foreign Ministry, a topic the two men had discussed throughout the summer. During the tense days surrounding the May crisis, when Ribbentrop had urged war against the Czechs, Weizsäcker had said, "I absolutely disagree with you!" and had pleaded his case. Confronted by his subordinate's objections, the foreign minister had immediately changed his mind and supported the "political disintegration" rather than the immediate invasion of Czechoslovakia.

However, as Ribbentrop realized that Hitler was determined on war, he hastened to fall into line. According to Erich Kordt, Ribbentrop's method of doing foreign policy research was to linger at the Reich Chancellery, trying to learn from hangers-on what Hitler was thinking. When he found out the führer's thoughts he fed them back to Hitler as his own ideas. But "if Ribbentrop found that Hitler had taken a stand different from what he had expected, he would immediately change his attitude."[33]

Ribbentrop now reminded Weizsäcker that the Foreign Ministry "must speak in confident and strong tones" about the Czech problem. If the Western Powers intervened, Ribbentrop asserted, Germany would fight and win. "The French would be beaten decisively in a great battle on the western frontier of Germany. And we are prepared for a war of any length, since we were well provided with raw materials and Göring was having so many planes built that we should be superior to any opponent."[34]

Weizsäcker demurred, saying that if war came, he did not believe that Germany could win it. He thought that it was possible to defeat another country only by occupying it or by starving it to death. To think that it was possible to win a modern war merely with airplanes, Weizsäcker said, "was a Utopian idea." Furthermore he did not believe that Germany had the endurance for a long war. "Ribbentrop then half acquiesced."

In his growing opposition to Hitler's impending war, Weizsäcker was moving along a path parallel to that being trod by General Beck at the General Staff.

Neville Chamberlain rose from the government front bench to address the House of Commons just before it adjourned for its summer recess. As the prime minister surveyed his realm he was feeling confident, not just about his political control of the House of Commons, but also about his increasing control over what was happening in Czechoslovakia.

As Europe moved further away from the dangerous May crisis, Chamberlain believed that the Sudeten issue would be worked out peacefully. In the wake of the Wiedemann visit and the proposed visit by Göring, Chamberlain's spirits soared. He wrote to his sister that "this is the most encouraging news from Berlin that I have heard yet and I hope signifies that they [the Germans] mean to behave respectably for the present."[35]

He now addressed the House with soothing words about the Sudeten issue. "I believe we all feel that the atmosphere is lighter, and that throughout the Continent there is a relaxation of that state of tension which six months ago was present."[36] The source of the prime minister's optimism was the revelation that he had just gotten the Czech government and the Sudeten Germans to agree to a British mediator. In fact the search for such a person had been going on since the May crisis had lit a fire under Chamberlain and Halifax. At the cabinet meeting of June 22, Halifax had told his colleagues that if negotiations between the Czechs and Sudeten Germans stalled, "he proposed to have a wise British subject available to slip off quickly to Central Europe to try and get the parties together again."[37] After going over a shortlist of names compiled by Sir Horace Wilson, the "wise British subject" Chamberlain and Halifax came up with was Lord Runciman, a sixty-eight-year-old shipping magnate who had served in the cabinet as president of the Board of Trade from 1931 to 1937. Sir Horace Wilson had privately described Runciman as having "a puzzling demeanor which might, in certain circumstances, be of advantage." He added: "Someone would have to accompany him and do most of the work, but he could be relied on to put the results across."[38]

Chamberlain assured the House that Lord Runciman was going out "in response to a request from the Czechoslovak government." He went not as an arbitrator but as "an investigator and mediator," who "would, of course, be independent of His Majesty's Government—in fact, he would be independent of all Governments" and "would act only in his personal capacity."[39]

All of these statements were false. Runciman had been foisted on the Czechs, not sent at their request. He might be called a mediator, but he knew that his real job was to pressure the Czechs to knuckle under to the Sudeten demands. Finally Runciman had no status as a private person; it was clear to everyone in Europe that he had influence *only* insofar as he *did* speak for the British government. Chamberlain asserted that he was independent in order to keep open the option of repudiating Runciman's recommendations if he found them uncongenial.

From the British Embassy in Berlin, Nevile Henderson commented for all the appeasers in His Majesty's Government: "I do not envy Lord Runciman the difficult and thankless job which he is undertaking. The Czechs are a pig-headed race and [President] Beneš not the least pig-headed among them."[40] Runciman might be phlegmatic, of puzzling demeanor, and in need of assistance, but Chamberlain and Halifax believed that he was reliable. They knew that their old colleague could discern the real agenda behind his assignment without needing to be told. For all its fragility the Runciman mission became Chamberlain's best hope for solving the Sudeten crisis in the late summer of 1938.

JULY 29, BERLIN, ARMY HEADQUARTERS, BENDLERSTRASSE

Beck had been alarmed to hear of Hitler's dismissal of Wiedemann's message from London. It was still another sign that the führer would not be dissuaded from aggression no matter what the British said or did.[41]

Beck believed that it was madness for Hitler to count on British noninvolvement in the event of war. He wrote at this time, "I think that it is a dangerous error to believe that Britain cannot wage a long war. The war effort of Britain has always been long-term, because her strength lies in the immeasurable resources of the Empire." Furthermore, if Britain entered the war, Beck thought, it would not be so much to save Czechoslovakia, but to defeat Germany, "the disturber of the peace," and to reestablish the principles of statesmanship that the British have always recognized: "Law, Christianity, Tolerance."[42] Beck's ability to see how German aggression might be perceived in Britain was typical of his restless, unconventional military mind.

Beck marched into Brauchitsch's office and issued an ultimatum. He must seek out Hitler and give him the following message: "The Commander-in-Chief of the Army, along with his most senior commanding generals, regret that they cannot assume responsibility for the conduct of a war of this nature without carrying a share of the guilt for it in the face of the people and of history. Should the Führer, therefore, insist on the prosecution of this war, they hereby resign from their posts."[43] As a result of his closed-door conversations with Oster, and his own growing certainty about Hitler's intentions, Beck had come to the startling idea of a generals' strike in the event of war!

Beck thought that if the generals refused to lead their troops, there could be no war. How could he publicize this idea? A generals' strike would work only if it had nearly unanimous support. Brauchitsch and Beck resolved to float this idea as soon as possible.

But Beck also gave some thought to more serious measures. He understood that a generals' strike might become the prelude to a putsch. "It will thereafter be necessary for the Army to be prepared, not only for possible war, but also for upheaval at home which it should be possible to confine to Berlin. Issue orders accordingly. Get Witzleben together with Helldorf," he wrote. The reference to the latter two was ominous. General Erwin von Witzleben commanded the III Corps and Wehrkreis III (Berlin), while Wolf von Helldorf was the police president of Berlin.[44] With Oster as his guide, Beck was now on the brink of embracing a full-scale revolt against Hitler and the Nazi regime.

July 30, Berlin, Abwehr Headquarters

Rear Admiral Canaris sat in his office reading the mass of memorandums, orders, and directives that were normally read by Lieutenant General Keitel, Hitler's chief of staff. Keitel had gone on vacation and asked Canaris to fill in for him. While Oster nudged Beck toward revolution, Canaris hesitated. He took the view that Hitler was bluffing President Beneš to secure a better deal for the Sudeten Germans, a belief that was widely shared among German officers. Despite the führer's saber rattling, Canaris did not think that he actually intended to go to war against the Czechs. Now, as he reviewed the papers in front of him, he became convinced that he had been wrong. There was no question that the Wehrmacht would march. Canaris was particularly impressed by the closely reasoned protests of General Beck.[45]

Throughout the spring and early summer, Canaris had wavered in his support for Oster's plan for revolt. No longer. Canaris was now convinced that Hitler was intent on war, and that he must be stopped.

Late July, Berlin, the Casino Club

Ian Colvin was a well-connected British journalist whose membership in Berlin's prestigious Casino Club opened many doors for him. One of his closest German friends was the old conservative anti-Nazi Ewald von Kleist-Schmenzin. Kleist, who was also very close to both Canaris and Oster, had come to Berlin from his Pomeranian estate to offer his help to his friends in the resistance. The British were on his mind. Colvin remembered that Kleist asked him if Britain would fight if Germany attacked Czechoslovakia.

Colvin answered, "I believe so. Perhaps only by a blockade at the outset." Kleist agreed, and then he dropped his voice and whispered: "The Admiral [Canaris] wants someone to go to London and find out. We have an offer to make the British and a warning to give them."[46]

JULY 31, BERLIN, ARMY HEADQUARTERS

Still smarting from his confrontation with Hitler in June, Colonel General Adam had been searching for a way to deter Hitler from going to war. One Sunday he met with General of Artillery Halder, Beck's second in command on the General Staff. They spoke of the dangerous situation that Hitler had created for Germany. Halder intimated that he had contacts with members of the British cabinet, freely acknowledging that this constituted "treasonous behavior." Adam replied: "If you seriously have such connections, then you must do everything you can to activate them. See to it that a person in authority is sent here to take some serious whacks at Hitler's knuckles. Maybe this will knock some sense into him." But Adam clearly thought that such a change of heart on the part of the führer was highly unlikely. When Halder said, "If Witzleben begins an attack, then the commanders in other parts of the Reich will have to do the same," Adam answered, "Let's go, I'm ready." Although he was still unclear about precisely who was behind the impending coup, Adam pledged his support to Halder.[47]

At the conclusion of the meeting, Halder asked Adam to sound out Colonel General Rundstedt on the idea of resistance to Hitler. Adam did so immediately, but found Rundstedt unreceptive, taking a "Hitler-knows-best attitude." After reporting back to Halder, Adam noted, "I never heard of those plans again."[48]

That may be, but Halder knew that it was unnecessary to keep the fiery general informed of detailed developments: When Oster and the General Staff officers decided that the moment for action had come, they knew that they could count on General Adam.

For the first and last time in the history of the Third Reich, the *Generalität* met without having been summoned by the führer. The conveners of this extraordinary meeting were a determined Beck and a reluctant Brauchitsch.

The first item of business was Beck's reading of his memorandum of July 16. Beck demanded that Hitler take into account [*berücksichtigen*] the concerns of the generals about the limits of German military strength and the request "to represent the unanimous view [*einheitliche Auffassung*] of the army." He set his cautionary remarks in the context of the geopolitical situation. An attack on Czechoslovakia would provoke the French and then the British to intervene. The French could bring overwhelming force to bear on Germany's western front within five days of mobilization. Hovering in the background, ready to join the war, would be Russia and the United States. (All the generals attending the meeting would have remembered that it was the arrival of the American Expeditionary Forces in great numbers in 1918 that had led to the German army's defeat in World War I.) Germany would be joined only by a few "insecure and insignificant" allies and would face insurmountable odds and certain defeat.[49]

Brauchitsch invited General Adam, commander of the western front, to comment. The crusty old soldier called Hitler's plan "a war of desperation beyond all hope." He said that the West Wall would be one-third finished in three to four months, but even then would be virtually useless. With the five regular and four reserve divisions that he had to defend the west, Adam predicted that his position would be quickly overrun. "I paint a black picture," he said somberly, "but it is the truth."[50]

Not one general present at the meeting disputed these estimates of the military situation. Only Generals Walther von Reichenau and Ernst Busch (commander of the VIII Corps) rebuked the others for their opposition to Hitler. But, according to Major General Maximilian Weichs, one of the participants, the consensus of the meeting was "that the mood of the people and the Army was generally against war, and that the organization, training, and equipment of the troops would be sufficient for a war against Czechoslovakia, but not against the European great powers"—the same conclusion that Beck had reached after the war game in June.[51]

Adam remembered that at the end of the meeting either General Kluge or General Liebmann said to him, "Everyone who is wearing two stars on his epaulette should resign out of protest," to which Adam replied, "That's exactly what I think—I will be interested in seeing how many will do it."[52]

This meeting had no definitive outcome, nor did Beck necessarily expect one. He wanted to inform his fellow commanders of the führer's folly and to set the stage for a concerted action if Hitler pressed forward with his plans. He told Erich Kordt that if Hitler insisted on his planned attack on Czechoslovakia there would be a generals' strike, and if Hitler tried "to force his way a revolt would take place."[53]

Meanwhile Reichenau scuttled off to tell Hitler about his mutinous generals.[54]

EARLY AUGUST, GREAT FOSTERS, OUTSIDE LONDON

Erich Kordt again flew to England, this time to meet with his brother at Weizsäcker's suggestion. It was ostensibly a vacation, or at least that's what Erich claimed. He had been working fourteen- to sixteen-hour days at the Foreign Ministry, and Ribbentrop had agreed to give him a few days off "because for the rest of August and September there wouldn't be time for it."

Erich suggested to Theo that they meet at Great Fosters, a Tudor castle near London that had been converted to a resort hotel. As they met in the lobby Theo remarked, "You've picked the right place. I just ran into Negus the Abyssinian, one of Mussolini's victims. A bad omen!" Erich noted that the cause of peace didn't look too good to him either. Omens notwithstanding, for a few days the brothers took walks and talked about diplomatic possibilities.

On his return to London, Erich met with his friend Philip Conwell-Evans at the Travellers' Club. The Englishman didn't seem

overly worried about the possibility of war. "Hitler cannot start a war
because of the complaints of the Germans in Czechoslovakia anymore," Conwell-Evans said. "We now have Lord Runciman in Prague
as the arbitrator between the Czech government and the Sudeten Germans. He'll bring about a compromise." Erich was incredulous at his
friend's complacency.

He told Conwell-Evans that he "should not labor under an illu-
sion. The situation was too serious and too dangerous. The previous
steps of the British government have had no influence on Hitler."
Conwell-Evans was thoughtful but reiterated that he believed Erich's
fears were exaggerated.[55] Erich knew that the Runciman mission was
doomed to failure. Hitler would never let Henlein accept a Czech
offer, no matter how generous it was. What would it take, he wondered, to get the British to see that?

AUGUST 10, THE BERGHOF

When Hitler heard about Beck's and Brauchitsch's meeting with the
Generalität, he was furious and resolved to get rid of Beck. He wanted
to eradicate the virus of Beck's "defeatism" before it could infect the
entire officer corps. Hitler hastily convened a meeting of some twenty
younger generals at the Berghof, just six days after all the generals had
heard Beck's and Adam's dire analyses. Hitler decided to concentrate
on younger generals who, at the outbreak of war, would be the chiefs
of staff of his field commanders, and who might also benefit from the
career opportunities that every war inevitably brings.

From the top of the mountain the generals could look down on
Salzburg, so recently part of an independent state, now incorporated
into the German Reich. The speech was vintage Hitler. He spoke for
three hours, explaining how the mounting injustices of the Czechs
toward their German minority made it imperative for the Reich to act
immediately. The führer discounted the Soviets, who were in no shape
to fight, and the British and the French, who lacked the will to do so.

Then Hitler made a tactical error. Pleased with his speech, and assuming that the young generals had been persuaded by his rhetoric, he opened the meeting to questions. General Gustav von Wietersheim, the senior officer present and General Adam's subordinate, repeated Adam's opinion that the West Wall could not withstand a French attack. Hitler, enraged, screamed at him, "I assure you, Herr General, that the line will hold not for three weeks but for three years!"[56] The discussion was over.

In the aftermath of the meeting, General Alfred Jodl, chief of the operations staff of the Army High Command (OKW), commented on recent disagreements between Hitler and his commanders: "Since water flows downhill, this defeatism may not only cause immense political damage, for the opposition between the generals' opinions and those of the führer is common talk, but may also constitute a danger for the morale of the troops."[57] Even Hitler's military sycophants were worried about their fellow generals' obvious lack of enthusiasm for the coming campaign against Czechoslovakia.

AUGUST 15, JÜTERBOG, THE ARTILLERY GROUNDS,
SOUTH OF BERLIN

General of Infantry Curt Liebmann, commander of the *Kriegsakademie*, along with the Wehrmacht's other leading generals, gathered at the artillery grounds for a demonstration of how its 15 cm howitzers could easily destroy concrete bunkers of the type the Czechs had constructed on their frontiers. The hypothetical attack "was treated as child's play, with the aim of encouraging the generals." Beck was beside himself with anger because of the misleading demonstration.[58]

After a day of observing these field maneuvers, the generals assembled to hear their führer. As Beck and Brauchitsch had predicted, Hitler delivered his justification for the coming attack on Czechoslovakia, referring to that country as "a Soviet Russian aircraft carrier." He reminded them that he had always been successful in his moves to restore Germany to its position of predominance in Europe. The führer assured his generals once again that France was "psychologically" unready for war and England was unprepared. When a questioner pressed him further about the possibility of French and British intervention, he answered, "Gentlemen, with that possibility you need not concern yourselves." Finally Hitler predicted that "by the end of the year we will be looking back at great success."

Although Liebmann was aware of Hitler's motives in calling the meeting—and dubious about them—he admitted that "Hitler's presentation was done with great self-assurance and we were more or less convinced that there would be a quick and victorious course for the planned enterprise." Despite the debacle of his meeting five days earlier with the younger generals, this meeting with the *Generalität* had been a triumph for the führer.

Beck immediately requested a conference with Brauchitsch, who this time refused to see him; he had had enough of Beck's hectoring dissent. When Beck finally managed to see Brauchitsch on August 18, he submitted his resignation. Beck urged Brauchitsch to join him in resigning, but Brauchitsch "hitched his collar a notch higher and said, 'I am a soldier; it is my duty to obey.' "[59]

Beck confided to Liebmann about his decision to resign, "If I wanted to preserve even one spark of self-respect, I could not have acted otherwise."[60] Several months later Beck reflected on his decision again. "When all was said and done," he said, "I sat in the seat of Moltke and Schlieffen and had an inheritance to administer. I could not quietly observe how this band of criminals let loose a war." After Brauchitsch told him that Hitler had rejected his memorandums and was committed to war against the Czechs, "I declared that if it came to war, I would do my duty as a gunner, but would not make myself available as chief of staff."[61]

Beck's intention in resigning, according to Gisevius, "was to give the signal for a general defection from Hitler." Oster supported his decision to resign. He thought that Beck's resignation would "make a lasting impression on the generals" and bring about the confrontation between the army and Hitler that Oster had tried to provoke during the Fritsch crisis.[62]

In mid-August, just after Hitler's speech at Jüterbog to his generals, Hans-Bernd Gisevius went to Düsseldorf to visit Carl Schmid, the head of regional government, and "one of our most reliable advisers on the 'inside.'" He wanted to find out what was happening out in the provinces from a man whom he deeply respected. Gisevius was especially interested in public opinion about the growing Sudeten crisis. Did Hitler have popular support for a military confrontation?

Schmid received Gisevius in his office and confided that Düsseldorfers were so horrified by the prospect of war that they were trying to convince themselves that Hitler was only bluffing, and that the führer shared their passionate desire for peace. "All remembered the horrors of the First World War; they feared those perils to the marrow of their bones," Schmid told Gisevius. "Everyone realized that war was a deadly serious matter. War meant bread cards, starvation rations, and intensified terrorism. War meant hundreds of thousands of lives, not to mention air raids. War meant a plunge into the abyss."[63] Believing that Hitler was not bluffing, Gisevius became more convinced than ever that Hitler had lost his unerring sense of what the people of Germany would tolerate. "He had strayed farther than ever before from the tenor of public opinion," Gisevius concluded.

While visiting Düsseldorf, Gisevius received an invitation from General of Artillery Hans Günther von Kluge, commander of Wehrkreis VI in nearby Münster, who sent a car to pick him up. Kluge wanted to hear the news from Berlin. Kluge wondered if Hitler could possibly be serious. He had heard Hitler's speech at Jüterbog but had dismissed it as mere boasting. Would Hitler actually plunge Germany into this unwinnable war? Gisevius was struck by Kluge's parting words and described them in his memoir. "As I shook hands with him for the last time on this earth, he bade me good-bye with a whimsical melancholy unusual in so high-ranking a general. 'Well,' he said, 'see you again in a mass grave.'"

Gisevius would have reported on this trip to Oster, who must have been heartened by this further evidence that popular opinion was decidedly against war, and that another Wehrkreis commander shared their views.

August 17, London, Park Lane Hotel

The majestic Park Lane Hotel, in London's elegant Mayfair, had always catered to wealthy and discreet travelers. Discretion was uppermost in the mind of Ewald von Kleist-Schmenzin as he checked in under an assumed name on the evening of August 17, 1938. Kleist had come to London on a secret mission at the request of Hans Oster and Beck.[64]

By late August, according to Fabian von Schlabrendorff, Oster's friend and later fellow conspirator, Oster had become the most important advocate among the conspirators for making contact with the British. "It became increasingly the concern of a number of members within the opposition that contacts with foreign countries be strengthened and relations with Britain improved in the hope of getting a chance to impress upon the British government the urgency of a firm stand against Hitler." Oster, Schlabrendorff noted, "was particularly active in trying to bring this about."[65]

As the Czech crisis deepened, and it became clear that the British would play a pivotal role, Oster called on Kleist for this delicate assignment. Summoning the Prussian aristocrat to Abwehr Headquarters, he told Kleist that the conspirators wanted assurances from the British that they would stand firm against Hitler in the event of an invasion of Czechoslovakia. Beck, now ready to revolt against the regime, said grimly to Kleist, "Bring me certain proof that England will fight if Czechoslovakia is attacked and I will make an end of this regime." Kleist asked what Beck would regard as proof. The general replied, "An open pledge to assist Czechoslovakia in the event of war."[66]

Kleist, a known enemy of the regime, had been fortunate in making his departure from Tempelhof Airport without the usual checks at customs and currency control. His cousin, a general, had appeared with a military car and had whisked him directly to the tarmac in front of the Junker 52 aircraft. With General Kleist running interference, the conspirator had escaped the authorities' scrutiny. Not until he was airborne did Kleist finally relax.[67]

Kleist asked his friend Ian Colvin to make arrangements for his trip in London.[68] Colvin, knowing that Kleist would be less conspicuous in the anonymity of a large hotel, suggested that he stay at the Park Lane. After arriving, Kleist was met by Lord Lloyd, an anti-appeasement politician, who took him to dinner in private rooms at Claridge's. Lloyd spoke no German, and Kleist spoke no English, but they managed tolerably well in French.

"Everything is decided, Lord Lloyd," Kleist told him excitedly. "The mobilisation plans are complete, zero day is fixed, and the Army group commanders have their orders. All will run according to plan at the end of September, and no one can stop it unless Britain speaks an open warning to Herr Hitler."[69] Lord Lloyd reported on his meeting with Kleist to Halifax.

AUGUST 18, LONDON, FOREIGN OFFICE

After breakfasting at his hotel, Kleist met with Sir Robert Vansittart. Vansittart had been briefed by Colvin about Kleist, and agreed to meet him at the Foreign Office. Kleist told Vansittart that war over Czechoslovakia was now "a complete certainty," and that it would take place on or before September 27, a date British officials had heard from other sources. In spite of the alarming nature of Kleist's pronouncement, Vansittart, like many British diplomats and politicians, refused to take Hitler's threat "at its ugly face value." He believed that "the appropriate answer to Germany's increasingly aggressive revisionism and, later, expansionism was a conciliatory policy of 'coming to terms' which would minimize the radicals' influence on Hitler."[70]

Vansittart, despite his legendary anti-German attitude, took a position on Hitler that was not much different from Chamberlain's. He believed that German politicians could be divided neatly into categories of "moderate" or "extreme." In this prevailing misconception,

Hitler was a relatively passive leader who floated between the two factions that were grappling for control of the Third Reich. Vansittart asked Kleist if Hitler was being "carried along" by the extremists. Kleist replied emphatically: "No, I do not mean that. There is only one real extremist and that is Hitler himself. He is the great danger and he is doing this entirely on his own."[71]

Kleist had a recommendation for the British government if it wished to stop Hitler—make an unequivocal statement of Britain's intention to defend Czechoslovakia. At the moment, Kleist said, Hitler was convinced that the British and the French had been bluffing in May, and that they intended to desert the Czechs. Kleist claimed that "a great part of the country [Germany] is sick of the present regime and even a part that is not sick of it is terribly alarmed at the prospect of war." He urged that some senior British statesman "should make a speech which would appeal to this element in Germany, emphasizing the horrors of war and the inevitable general catastrophe to which it would lead." Kleist emphasized that he had come at great risk to himself, "with a rope around [my] neck." However, he and his friends in the army were willing to move against the regime, but that "alone they could do nothing without assistance from outside on the lines he had suggested."[72] Never before had any resister been so candid with an outsider about their intentions. The time for caution had passed; the conspirators knew that Hitler's time bomb was ticking.

Vansittart passed on the report of his meeting with Kleist to Halifax, who in turn sent it to Chamberlain.[73]

AUGUST 19, CHARTWELL, WESTERHAM, KENT

Kleist then traveled to meet with Winston Churchill at his country house, Chartwell. Churchill had been out of office since 1929, and had broken with the Conservative Party leadership over the issue of Indian independence—he was against it—in 1931. In the summer of 1938, as the Czech crisis deepened, Churchill had given speeches, written newspaper articles, and sent notes to ministers arguing that His Majesty's

Government should take a strong stand against Hitler's aggression. Specifically, he advocated that the government should fashion a "Grand Alliance"—to include Britain, France, the Soviet Union, and perhaps even the United States—to guarantee Czech independence.

Churchill, reputedly the worst driver in England, drove Kleist around the Chartwell estate and the surrounding countryside in his automobile while his son, Randolph, took notes of their conversation. Kleist stressed that the generals opposed war, believing that an attack on the Czechs would inevitably bring in France and Britain, and that Germany could not hold out for more than three months in such a conflict. With encouragement, Kleist thought, the generals "might refuse to march." They alone had the power at this late date to stop the war. Kleist appealed to Churchill, as he had to Vansittart, for some gesture "to crystallize the wide-spread and indeed, universal anti-war sentiment in Germany."[74]

Churchill knew that war was approaching but that ordinary Britons had been lulled into a dreamy sense of security by Chamberlain's reassuring rhetoric. Churchill's repeated alarms about German rearmament had fallen on deaf ears, both inside the government and in the country as a whole. He understood the importance of Kleist's offer, particularly at this dark moment. In Kleist, the British government had an invaluable ally—a German anti-Nazi who had hinted broadly that some German generals were ready to take action against the Nazi regime and perhaps remove the terrible threat of war. Churchill assured Kleist that he would do what he could.

AUGUST 19, BERLIN, FOREIGN MINISTRY

Ribbentrop called in Weizsäcker and told him that Hitler had decided to eliminate the Czechs by force. Weizsäcker tried to object, but Ribbentrop cut him off, explaining that Hitler had "never made a mistake." The führer had made decisions, like the occupation of the Rhineland, which had been questioned at the time but had proved to be correct. Weizsäcker remembered that Ribbentrop told him, "I ought to have faith in his genius, just as he had as a result of many years' experience." Ribbentrop urged Weizsäcker to embrace this "blind faith" in the führer. If he was unable to do so, Ribbentrop said ominously, he "should regret it afterwards."

Weizsäcker went over to the conspiracy. Speaking of Hitler and the Nazis, he said, "Somehow and sometime or other we should have to get rid of them, and meanwhile we should have to put a spoke in the wheel of the cart as it rolled towards the abyss." He added ominously, "If we could not use the normal means to this end, we should have to work in secret."

Weizsäcker began conferring with General Beck, who shared his opinion that "a European war must be prevented." Even after Beck had resigned, Weizsäcker stayed in touch with him through mutual contacts in the Abwehr. Weizsäcker was pleased to discover that Beck's successor, General of Artillery Franz Halder, shared his views.[75]

If Hitler had known that highly placed dissenters from the Foreign Ministry, Abwehr, the General Staff, and the Army High Command were now in secret communication with one another, it might have dimmed his bright optimism about the likely success of Plan Green.

AUGUST 20, CHEQUERS, BUCKINGHAMSHIRE

From Chequers, the country estate of British prime ministers, Chamberlain reviewed the report on Kleist and then quickly wrote to Halifax: "I take it that Von Kleist is violently anti-Hitler and is extremely anxious to stir up his friends in Germany to make an attempt at his overthrow. He reminds me of the Jacobites [British traitors] at the Court of France in King William's time and I think that we must discount a good deal of what he says."[76] Despite Kleist's proffered intelligence about the date of the attack and his information about resistance to Hitler within the army, which reinforced information that the government had received from other sources, he was, in Chamberlain's view, a traitor pure and simple. For Chamberlain, Kleist's message held no significance. At Halifax's prodding Chamberlain reluctantly agreed to recall the British ambassador, Sir Nevile Henderson, to London for consultations, a mild form of diplomatic protest. But if he thought that this timid action would give Hitler pause, Chamberlain was sadly mistaken. There is no indication that the führer even noticed that Henderson had left Berlin or, if he did, that he cared.

The icy reception given to the conspirators by Chamberlain is regrettable but understandable. The conspirators had interjected themselves at a most awkward time—when Chamberlain was already planning personal diplomacy that he believed would solve all conflicts between Britain and Germany with one dramatic gesture.

AUGUST 20, LONDON, FOREIGN OFFICE

Sir Robert Vansittart pondered the question of how much, if any, credence he should give to the German plotters. Despite his antipathy to Hitler, he wondered whether there was really a significant difference between Nazism and the old-fashioned German nationalism of Kleist and Goerdeler. When Kleist, like Goerdeler before him, sketched out a plan for a post–Hitler Germany, it included revisions of the territorial boundaries imposed by the postwar treaties. Kleist was particularly anxious to restore German territories in the east, while Goerdeler spoke warmly about a "Greater Germany" that would include Austrians and perhaps even Sudeten Germans. They, and other emissaries, often supported the restoration of a monarchy.

Viewing Goerdeler and Kleist with extreme suspicion, Vansittart concluded that there was really very little difference between the conspirators and the Nazis. "The same sort of ambitions are sponsored by a different body of men, and that is about all."[77]

At the root of this conclusion was the Foreign Office's continued belief that Hitler was a passive character surrounded by factions of "moderates" and "extremists." This misperception, in its own way, was as tragic as Chamberlain's dismissal of the conspirators as a gang of traitors. Even Kleist's blunt assessment of the führer ("There is only one extremist and that is Hitler himself") failed to change the Foreign

Office view of Hitler that had been its standard assessment for nearly two decades.[78] By focusing on the similarities between the programs of the nationalist conspirators and Hitler, Vansittart and his colleagues ignored the obvious and vital difference between them: Hitler was a violent man who meant to impose his will on Europe by force, whereas the conspirators embraced the "Spirit of Locarno" that had prevailed in Europe in the late 1920s. They wanted Germany restored to its former greatness, but they understood that this could be achieved only through peaceful negotiation.

Vansittart and his colleagues did not understand that German conservatives and nationalists might be moral and religious men who were appalled at the lawlessness, brutality, and inhumanity of the Nazis. It was a profound failure of analysis that led Vansittart and the Foreign Office to dismiss Kleist and his message so cavalierly.

AUGUST 20, CHARTWELL, WESTERHAM, KENT

Churchill's response to his meeting with Kleist was altogether different from Vansittart's. He urged Halifax to take the conspirators seriously, predicting that their planned coup would bring "a new system of government within 48 hours. Such a government, probably of a monarchist character, could guarantee stability and end the fear of war for ever."[79] Churchill enthusiastically wrote the letter that Kleist had requested, although both he and Kleist knew that as a parliamentary backbencher he could in no way make binding commitments for his government. Churchill wrote that an armed attack by Germany on Czechoslovakia "will rouse the whole British Empire and compel the gravest decisions." Presciently, he went on to write "such a war, once started, would be fought to the bitter end" and would likely last for years, not months. Finally he warned that not even the vaunted German air superiority would guarantee a quick victory: "All the great Nations engaged in the struggle, once started, would fight on for victory or death."[80]

While Chamberlain and most of the cabinet shared the delusion of appeasement, and Hitler dreamed of a conflict that would be short and victorious, Churchill, Oster, and the German conspirators foresaw that a renewed world war would unleash a prolonged hell on earth.

AUGUST 21, BERLIN, REICH CHANCELLERY

Hitler announced that he would accept the resignation of the chief of the General Staff that had been proffered several days earlier. The führer's antipathy to Beck had been building throughout the year. In the wake of the Fritsch case, he had said to Franz Gürtner, minister of justice, "The only man I fear is Beck. That man would be capable of acting against me."[81] Hitler clearly understood the difference between himself and Beck and, by extension, the growing difference between himself and many of his generals when he said, "Beck was hesitant, he put caution above everything else . . . [He] wanted to build up this or that, but not for the enlargement or strengthening of the Fatherland but only for its security."[82]

For Hitler relief replaced fear and anger. He had at last rid himself of this meddlesome chief of the General Staff. He also insisted that Beck not make his resignation public "on the pretext that the foreign situation was already tense enough." According to Gisevius, Beck did not argue with this decision. "With exaggerated aristocratic sensitivity, he resigned without making a fuss about it"—which undercut one of his initial purposes for resigning! Even Erich Kordt said, "General Beck played his cards too soon and was removed with considerable loss of face."[83] Beck, who had been the rallying point for the military opposition, and who had so recently taken the extraordinary step of bruiting the idea of a generals' strike, failed to see the political opportunity in publicizing his resignation. "Thus," Gisevius lamented, "the German people heard not a word about the significant action of their chief of the general staff."[84]

Beck's successor, Franz Halder, never the soul of tact, chastised him for his decision: " 'Now you see,' he said, 'what one can achieve with intellectual memoranda and elegant gestures of resignation. The time for memoranda is past. We must adopt other methods.' Resigned and bitter, Beck answered, 'I should like to believe that you are right. But now everything depends on you.' "[85]

Why did Beck resign? It had been his responsibility to think globally on behalf of the state. He had done so and reported his conclusions faithfully to Hitler, but he had been ignored. Now, he told Hossbach, a war begun by the führer and lost would have much graver consequences for Germany than even World War I. Beck specifically emphasized the issue of guilt. "The highest military authorities would not be acquitted of the political responsibility for the outbreak of war if they had done nothing to prevent it."[86] He refused to bear that burden.

Beck's somber journey from dissent to revolt was a prototype for other senior officers. They could begin to think about a putsch against Hitler only when they had realized that nothing else would stop the führer on his mad dash toward destruction. Beck, admired for his sharp mind and his unsullied integrity, was the general who commanded almost universal respect among his brother officers in 1938. His insistence on alerting his fellow generals to the political implications of Hitler's policies at the August 4 meeting had been crucial. With only Reichenau and Busch dissenting, the generals had accepted Beck's position, which Oster expected could be translated into support for action against Hitler if war seemed imminent.

Oster also learned lessons from his association with Beck in the spring and summer, and from Beck's resignation. Oster realized that Beck's hope for unified army action was illusory; the generals' strike, which depended on unanimity, would not work. Oster saw as well that Hitler was impervious to rational arguments: The only hope of saving the peace was a coup to eliminate him.[87] He knew that he must turn his attention to Beck's successor, General Halder, and through him, General Brauchitsch.

Although Beck's resignation took him out of the center of the opposition, over the weeks to come he hovered in the background, offering the resisters support and advice.[88]

Europe on the Edge of War:
Oster Brings the Conspiracy to Life

With Beck's resignation, Oster took deliberate control of the resistance movement. He knew that time was running out. Hitler had not budged from his deadline of October 1 for an invasion of Czechoslovakia. Fortunately Oster had extended his network widely. According to Gisevius, "We had our spies everywhere—in the war ministry, the police headquarters, the ministry of the interior, and especially in the foreign office. All the various threads came together in Oster's office; there we had our co-ordination center and there we would press the alarm button."[1]

Oster thought that Hitler might issue an ultimatum that would lead to an invasion of Czechoslovakia as early as his planned speech on September 12 at the annual Nazi Party rally in Nuremberg. Although he knew that he needed to act quickly, in the wake of Beck's resignation and Brauchitsch's indecision, no logistical planning had been done. The next few weeks would be crucial.

Meanwhile Halifax and Chamberlain were becoming alarmed.

Their hopes for the Runciman mission had gone up in smoke as Henlein, acting under Hitler's orders, had stubbornly refused all compromises proposed by the Czech government. Hitler kept the British politicians off balance with his inflammatory rhetoric and barely veiled threats. Parliament had not met since late June, and cabinet members had scattered across the Continent, enjoying their summer holidays while Europe lurched toward imminent war. It was against this background that Hans Oster began in earnest to construct the conspiracy against Hitler.

AUGUST 25, BERLIN, ARMY HEADQUARTERS

General of Infantry Liebmann had just returned from a five-day tour of his new command, the Fifth Army, deployed along the northern part of the West Wall facing Belgium. (Liebmann's superior for the entire western front was Colonel General Adam, commander of the army group that included the Fifth Army.) Appalled by the condition of the defenses in his sector, Liebmann had had conversations the previous day with Generals Jodl and Keitel, in which they had made light of his concerns.[2]

He now wrote a formal report about the defensibility of his assigned area. The forces under his command he deemed "wholly inadequate in numbers and quality." The fortifications in his area were incomplete, and "only a third" would be available by October 1. Liebmann noted that the most optimistic assessment of the West Wall was that it would act as a "bluff" that might forestall a British and French attack. "But if they attacked," he wrote, "one could count on the left bank of the Rhine being lost before the forces in Bohemia—the bulk of all available active divisions—could intervene." Liebmann never received a reply.

Oster counted for support on Liebmann and other generals who were increasingly alarmed by the state of Germany's defenses as war approached. He hoped that they would be the commanders who, rather than see Germany defeated, would turn their wrath on Hitler, particularly in support of a revolt led by the Army High Command.

AUGUST 27–29, HITLER'S TRAIN, AACHEN TO ISTEIN

Colonel General Adam received orders to report to Hitler's train to conduct the führer on an inspection tour of the western front. He found that the train was jammed with Hitler's swollen retinue and their aides. "There was hardly any room to spread out the maps," Adam groused.[3]

While Jodl enthused in his diary that the construction of such a strong defensive position in such a short period of time was "magnificent," Adam saw the situation differently.[4] He read his prepared remarks on the inadequacy of the West Wall defenses, but Hitler countered with facts and figures that Adam knew to be wildly inaccurate. Hitler said that "only a scoundrel could fail to hold that position." Indeed, he lamented to Adam that his other responsibilities precluded him taking his dream job. "I regret that I am the führer and Reich chancellor and cannot be the first commander of the western front!"[5]

"Once again," Adam noted, "I saw this man's lack of education, his inability when faced with reality, his lack of knowledge of foreigners and their mentality, and his mendacity." The mood quickly became tense as the two men verbally dueled, with Adam not giving an inch of ground. He left Hitler's railway car feeling a mixture of "triumph, anger and a willingness for further malice."

Adam and his fellow generals knew that all Hitler's blustering could not change the weakness of the West Wall. A handful of German divisions behind incomplete fortifications would prove to be a woefully inadequate defense against an all-out French attack. This nightmare of French divisions pouring into the heart of Germany virtually unopposed haunted the Wehrmacht's generals as war drew closer in the late summer of 1938.

For the remainder of the three-day tour Adam tried to avoid sitting next to Hitler whenever possible. When forced to join the führer at the table, Adam was disgusted by his uncouth table manners and his hectoring vegetarianism. Hitler self-righteously derided the company for eating "animal cadavers," Adam noted, while he dined on eggs stuffed with caviar! "Our mutual revulsion expressed itself in unpleasant ways."[6]

AUGUST 27, BERLIN, ARMY HEADQUARTERS

In one of his first acts in office, General of Artillery Halder, who now succeeded Beck as chief of the General Staff, summoned Oster. Halder wanted to know all the details and political implications of the Fritsch affair. Oster was still indignant at the army's having muffed so precious an opportunity for revolt. As their conversation progressed, Oster "was left with the impression . . . that Halder seriously intended to make up for past neglect."[7]

Halder came right to the point. He asked Oster what preparations had been made for a coup d'état. Oster answered that he had been unable to do anything thus far because of the resignation of Beck and the uncertainty of dealing with Brauchitsch. When Halder asked about potential conspirators among civilians, Oster mentioned Goerdeler and Schacht. Although Halder declined to meet the garrulous and indiscreet Goerdeler, he was anxious to discuss matters with the highly respected Schacht. Oster agreed to set up a meeting.

Oster's relationship with Halder was delicate. Although Oster was undoubtedly pleased that the new chief of the General Staff was involved in the conspiracy, he had reservations about the depth of Halder's commitment. When the moment for action came, could he count on him? In the interest of bringing the conspiracy to fruition, Oster knew that he would have to maintain control surreptitiously while seeming to cede leadership to Halder.[8]

AUGUST 27, SCOTLAND, LANARK

Sir John Simon, a member of the "inner cabinet" (Chamberlain and three high-ranking ministers), made a speech in Lanark in response to the government's increasing concern about the deteriorating situation in the Sudetenland. Simon's speech had been prompted by a report from Colonel Frank Mason-Macfarlane, Britain's military attaché in Berlin, which concluded, "We have to decide whether the moment is ripe to say 'No' or whether we are prepared to let Herr Hitler attempt what he hopes to be a rapid and possibly bloodless victory over the Czechs."[9] Simon did nothing more than reiterate the theme of a speech given in March by Chamberlain: If Germany attacked Czechoslovakia, and France honored its treaty obligations, "there was no telling what might happen." This bland policy statement of His Majesty's Government, unchanged since the Anschluss despite Hitler's increasingly violent threats toward Czechoslovakia, was as close to a warning as the British were willing to go.

AUGUST 27, BERLIN, ABWEHR HEADQUARTERS

Kleist strode into Abwehr Headquarters, announcing that he would report only to Admiral Canaris. His mission to England, he told the admiral, had been a failure: "I have found nobody in London who wishes to take this opportunity to wage a preventive war," said Kleist despondently. "I have the impression that they wish to avoid a war at almost any cost this year. Yet they may slip into it without wishing to. They say that it is not possible under the British constitution to commit themselves on a situation that has not arisen."[10] Churchill had written a letter, but both Canaris and Kleist knew that it had little value except as a morale booster.

AUGUST 27, BERLIN, ABWEHR HEADQUARTERS

In a meeting of Karl Hermann Frank, one of Henlein's deputies in the SDP; Admiral Canaris; and Lieutenant Colonel Helmuth Groscurth, Frank reported on his previous day's visit with the führer. Frank had come to Berlin for his marching orders. He revealed that Hitler had ordered him to provoke incidents in Czechoslovakia that would justify Germany's intervention. The führer had cursed Beneš and told Frank to bring the Czech president to him alive so that he could hang him personally. Canaris asked Frank directly to get Henlein to tell Hitler about his doubts concerning the war. Frank questioned the Abwehr officers' pessimism about Germany's chances in a world war, but he admitted that he also shared it.[11]

Apparently Hitler's aggressive intent toward Czechoslovakia alarmed even the puppets he had charged to carry out his will in the Sudetenland.

LATE AUGUST, LONDON, NUMBER 10 DOWNING STREET

Theo Kordt sought out Sir Horace Wilson at his office in Number 10. The two had met several times since July to discuss Czechoslovakia. In a two-hour conversation Theo laid out all the indications that Hitler intended to go to war over the Sudetenland. He said it was imperative "to destroy Hitler's hope of an easy victory by not having to fear intervention by Great Britain and France." He implored Sir Horace to persuade Chamberlain to support France if it fulfilled its treaty obligations to Czechoslovakia. If the French should become embroiled in the conflict, Kordt argued, Great Britain "cannot then stand aside."[12]

If either participant in this discussion noted the irony of a German diplomat urging a reluctant British politician to take a firm stand against an aggressive Germany, neither saw fit to comment on it.

AUGUST 28, BERLIN, ABWEHR HEADQUARTERS

The Abwehr had its offices in the massive Wehrmacht Headquarters complex, which today house Germany's Ministry of Defense. Situated on Berlin's Tirpitzufer (named for Grand Admiral Alfred von Tirpitz, the architect of the German navy prior to World War I; now Reichpietschufer), overlooking the chestnut trees on the Landwehrkanal, it was a short walk from the Foreign Ministry offices on Wilhelmstrasse, in the heart of the government district.[13] The two men meeting in the Abwehr Headquarters on a late August morning in 1938 couldn't have been more different in appearance and personality. The host, Lieutenant Colonel Hans Oster, was tall and slender, his handsome face dominated by penetrating gray eyes. Oster managed to look raffish even in his field gray Wehrmacht uniform. Men were attracted by his boisterous good humor and his charismatic personality; women were attracted by his physical elegance. His guest, Erich Kordt of the Foreign Ministry, wearing a rumpled suit, was a small and bespectacled diplomat with a round face that gave him an owlish appearance.

Kordt and Oster had been moving along parallel paths, trying to derail Hitler's invasion plans, but until this day they had never met. With the Czech crisis rapidly gathering steam, Oster had to find new allies who could carry out parts of the plan that was forming in his mind. He knew from Canaris that Erich Kordt and his brother, Theo, along with Weizsäcker, were among the sympathetic anti-Nazis in the Foreign Ministry. Despite the risk involved in contacting them, Oster had a delicate task for which he thought Kordt might be the appropriate person. He told Kordt about the outlines of the plot and then explained a key premise. The commander in chief of the army, Colonel General Brauchitsch, knew that a conspiracy had been forming among his subordinates, but he was ambivalent about joining it. "If Brauchitsch could be won over the plan must succeed," Oster confided in Kordt. "Brauchitsch is not a political person," Oster continued, "but he approved of Beck's memorandum, so he can't be in agreement with Hitler's military plans." Kordt wondered what all this had to do with him.[14]

Oster explained. He was concerned that Brauchitsch had been persuaded by Hitler's assurances that he knew the mind-set of the Western leaders, and that they would not intervene. He speculated that there was every possibility that Brauchitsch might go against the judgment of the other generals and support Hitler's plans for an invasion of Czechoslovakia.

"Make an appointment with Brauchitsch," Oster said, speaking more as if he were giving an order than making a request. "If somebody like you would give him a clear picture of the foreign political situation, perhaps it will impress him." Oster believed that Kordt, the highly placed diplomat, would have the credibility with the commander in chief in talking about the likely French and British response that an Abwehr officer would not. Oster had a few final words of advice. "Brauchitsch is reserved but is a respectable man. He will not betray you. But leave it up to him to draw conclusions from your presentation," Oster concluded with a smile. "Officers do not want to be taught by civilians."

Kordt welcomed the opportunity to join actively with the military resisters. He immediately agreed to carry out Oster's request. With Hitler's actions drawing Germany ever closer to a world war, Kordt and Oster began a collaboration that would soon bring them to the brink of a revolution against Hitler and the Nazi regime.

AUGUST 29, BERLIN, ARMY HEADQUARTERS

Erich Kordt immediately secured an appointment with Brauchitsch. As Oster had reminded him the previous day, it was essential for the conspirators to have Brauchitsch's full cooperation. Kordt had first sought permission from his superior, Ernst von Weizsäcker, to speak with Brauchitsch. The state secretary had urged him to do so, saying, "Make him understand everything."[15]

The general listened politely as Kordt explained the most recent developments in world politics as he saw them from the Foreign Ministry. He assured Brauchitsch that he had access to the same information as did Ribbentrop and Hitler, but that he had reached very different conclusions. In a potential conflict over Czechoslovakia, Kordt said, Germany would be left to fight without allies.

"And on what do Hitler and Ribbentrop base their hopes for victory?" Brauchitsch asked.

Kordt replied indirectly. "Not too long ago I read a circular from Ribbentrop to all the German foreign missions that had been agreed upon by Hitler." He summarized its content for Brauchitsch. "The Western Powers will not dare to spike Germany's guns if the Czech problem is solved with violence. But if they are still so blind as to offer resistance toward National Socialist Germany, then 75 million Germans will pounce on their enemies like one man and destroy them. The leadership will follow their chosen path coldly, quickly, and without deviating."

The general considered the implications of this directive. Then he asked, "What were your expectations in coming to me?"

Kordt replied, "In your hands, Herr General, lies the destiny of the German army and thus the destiny of the German people."

Brauchitsch thought for a moment and then said in a hoarse voice as he shook Kordt's hand, "I thank you for your message, which is of great value to me." Kordt stood for a moment, thinking that the commander in chief might add a clarifying word, but he did not. "He had pressed his thin lips together and said nothing more."

Kordt had hoped for a definitive commitment. He came away from his meeting disappointed, remembering only that he "could not get positive assurance that [Brauchitsch] would act."[16]

After the meeting Kordt thought about Brauchitsch's hesitation. How could the commander in chief refuse to act "if the destiny of the German nation was in danger?" Kordt wondered. The German people, Kordt knew, still had no understanding of Hitler's plans.

Kordt discussed the situation further with Canaris and Oster, and also asked them about Beck's and Halder's plans. They all agreed that without an open declaration by the British government of its willingness to intervene in the event of a German-Czech war, the German people would fail to see the danger confronting Germany, and thus would not understand the need for a coup. Without this popular support their planned coup could not succeed.

Still, Kordt took comfort in the expectation that the British, now having been apprised of Hitler's intentions by a succession of emissaries, would soon issue such a statement. Once the British statement was broadcast in Germany (it was not yet forbidden to listen to foreign radio stations), Kordt believed that "it would spread like wildfire. A revolutionary situation would develop as a prelude to a successful action against Hitler."[17]

Kordt understood that the vast majority of the German people did not want war, and that if it looked as though Hitler was leading them into one, his popular support could vanish overnight.

AUGUST 30, LONDON, NUMBER 10 DOWNING STREET

Throughout August, Lord Runciman had made no headway in recon-
ciling the differences between the Sudeten Germans and the Czechs.
Despite President Beneš's reluctant accommodation to their demands,
the Sudeten German leaders had rejected every offer. In Germany, the
situation also looked ominous. Other intelligence sources besides Kleist
had warned Halifax that Hitler intended to go to war against Czechoslovakia by the end of September. The German army was on extensive
maneuvers and had begun preliminary preparations for war. The mild
warning delivered by Sir John Simon on August 27 had made no
impression on the Germans. Reluctantly Chamberlain and Halifax
decided that they must recall their cabinet colleagues from their summer holidays and address the issue of what to do about the mounting
crisis over the Sudetenland.[18]

On August 30, with four cabinet members unable to reach London, the prime minister and the remaining seventeen members, along
with Ambassador Nevile Henderson, met in the majestic Cabinet
Room at Number 10 Downing Street.

The cabinet has sometimes been called "the Board of Directors of
the United Kingdom." It is an apt description. Cabinet decisions were
made collectively. While an issue was under discussion, each cabinet
member was expected to comment on it as candidly as he wished.
Once a consensus had been reached, however, a member who had disagreed with the decision had no choice but to support it or resign.

In its fifteen months of existence, the Chamberlain cabinet had
been reasonably free of internal strife. Its most visible defection had
been Anthony Eden, who had resigned as foreign minister in February
1938. By late August 1938, there were only a few cabinet members, led
by Alfred Duff Cooper, First Lord of the Admiralty, who regularly
expressed reservations about the Chamberlain/Halifax policy of
appeasement.

As Halifax entered the room on that warm summer day, he surveyed the grim faces of his colleagues. They were all aware of the circumstances that had brought them together for this emergency meeting.

Halifax knew that the agenda contained only one item—in fact, only one question: Should Great Britain formally warn Germany that an attack on Czechoslovakia would inevitably mean British intervention? After briefly reviewing recent events, Halifax laid out the policy alternatives. Underlying the British decision was an uncertainty about Hitler's real intentions in the Sudeten crisis. Was he determined to intervene by force, or—as Henderson believed—was he still undecided?

What could Britain do? Halifax said plainly, "The only deterrent which would be likely to be effective would be an announcement that if Germany invaded Czechoslovakia we should declare war upon her."[19] But what, Halifax asked, would be the implications of such an announcement? He thought that it would probably divide public opinion, both in Britain and in the empire. Halifax came to the nub of the issue when he said that he "would feel extremely uneasy at making any threat if he was not absolutely certain that the country would carry it out." Surely that statement resonated in the minds of the ministers around the table. They understood that it would be far more dangerous to make a threat and not carry it out when challenged than not to make the threat at all. That brought them to the issue of whether they were willing to go to war against Germany over Czechoslovakia.

To his credit Halifax acknowledged that there was more at stake than just the defense of Czechoslovakia. "We were, in effect, concerned with the attempt of the dictator countries to attain their ends by force." But Halifax immediately qualified this Churchillian statement of principle by asking "whether it was justifiable to fight a certain war in order to forestall a possible war later." In effect this question subtly reintroduced the issue of whether Hitler was really intent on using force.

Halifax was deeply pessimistic about the outcome of a war over Czechoslovakia. "There was nothing which we in this country or France, or Russia could do which would prevent Czechoslovakia from being overrun by Germany." (Ironically, this estimate of Germany's overwhelming military force was shared by neither Britain's military attaché in Prague nor Oster and his Abwehr colleagues. It was precisely because the latter and the army commanders believed that such an action would end in defeat that they had been able to recruit generals to the growing conspiracy.[20]) Halifax said that he thought it was "unlikely" that "any peace reached at the end of such a war would recreate Czechoslovakia as it existed today." Halifax was referring to the inclusion of dissident minorities—Hungarians and Poles as well as Germans—within a predominantly Czech state, rendering it inherently unstable.

Halifax's statement reverberated around the room. Not only had the foreign secretary said that a war to save Czechoslovakia was militarily unwinnable, but that the country could not be reconstituted even if the war *could* be won. While Halifax's analysis of international relations was more sophisticated than Chamberlain's, his conclusions were identical—Britain should not intervene under any circumstances. Halifax recommended taking a publicly ambiguous stance: "In effect we should try to keep Herr Hitler guessing."

Halifax briefly reviewed other options. Churchill had called for "a joint note to Berlin from a number of Powers." The flaw in that scheme, as Halifax saw it, was that their fellow signees "would probably ask embarrassing questions as to our attitude in the event of Germany invading Czechoslovakia."

Halifax was even more dismissive of overtures that he had received from "moderate Germans"—that is, Kleist and Goerdeler—who had promised that if His Majesty's Government took a strong stand against German aggression, "the Hitler regime would crack." Halifax noted that "he received these messages with some reserve." He "did not believe that the internal Regime of one country was destroyed as a result of action taken by some other countries." With this vacuous generalization, Halifax wrote off the pledge of the German conspirators to bring down the Nazi regime.

The prime minister spoke. Chamberlain also made reference to the conspirators, who had claimed that if Britain took a strong stand against Nazi aggression, the result "would probably be . . . a revolution which would upset Herr Hitler." The prime minister "always came back to the same conclusion reached by the Foreign Secretary." He, too, dismissed the conspirators' promises to overthrow the German government.

One by one, other ministers followed suit. Only Duff Cooper strenuously dissented. He noted that "there was a story going around that the German Generals had told Herr Hitler that they were not ready for war." He explained that a decisive diplomatic victory without having to force his doubting generals into war was precisely what Hitler wanted. Duff Cooper directly challenged Halifax's pessimistic assessment of the course of hostilities. "If war came," he believed, "Czechoslovakia would fight bravely and well and the French would go to their assistance." At the very least, he concluded, "he thought that we ought to show that we were thinking of the possibilities of using force." Duff Cooper's plea went unheeded.[21] There would be no unequivocal warning to Germany. His Majesty's Government affirmed Halifax's "guessing strategy" as its policy.

Behind this decision to stand pat in the face of Nazi preparations for war was the prime minister's hidden agenda. On August 30 Sir Horace Wilson, Chamberlain's confidant, made the first written record of the plan that would control the prime minister's actions over the next month: "Plan Z." It had first been discussed with Sir John Simon and Lord Halifax several days earlier. According to Sir Horace, "It is to come into operation only in certain circumstances. . . . The success of the plan, if it is to be put into operation, depends upon its being a complete surprise, and it is vital nothing should be said about it."[22]

In its inception Plan Z had Chamberlain flying to Germany to confer with Hitler just before a German attack on Czechoslovakia. To heighten the drama, Chamberlain had no intention of telling Hitler that he was coming. He would land in Germany—assuming, of course, that his plane was not shot down by the Luftwaffe—and suddenly restore peace to Europe. Chamberlain confided to his sister, "I keep racking my brains to try and devise some means of averting a catastrophe, if it should seem to be upon us. I thought of one so unconventional and daring that it rather took Halifax's breath away."[23] No doubt.

What can we suppose Halifax thought of his aged prime minister suddenly landing like a deus ex machina to stop German troops from invading Czechoslovakia? If Halifax cautioned him about it, there is no trace in the historical record. This White Knight fantasy came to dominate the reveries of Neville Chamberlain and to eclipse other diplomatic alternatives for dealing with the Sudeten crisis.

While in Berlin, Oster plotted to overthrow Hitler, in London, Chamberlain plotted to appease him.

AUGUST 31, CHARTWELL

Although he was not a member of the government, Churchill had excellent sources of information and sensitive political antennae. He either discovered or intuited what had gone on in the cabinet meeting the day before. From Chartwell he hurriedly penned a letter to Halifax, reiterating his support for a joint note from Britain, France, and Russia to Germany. Churchill suggested that the note should be given to President Roosevelt, who would then present it to the führer, with the added weight of the United States behind it. In contrast to most British politicians—especially Chamberlain—who had contempt for Americans, Churchill (whose mother, Jennie Jerome, was a New Yorker), admired and respected them. He also was sufficiently prescient to see that the United States would be Britain's most formidable potential ally in the event of another world war.

"The second step that might save the situation," Churchill opined, "would be fleet movements, and the placing of the reserve flotillas and cruiser squadrons into full commission."[24] Churchill knew that the Germans feared the Royal Navy, and that Hitler listened only to threats, not to pleas.

Churchill asked Halifax to show his letter to the prime minister, which he did. Chamberlain was unmoved.[25]

SEPTEMBER 1, BERLIN, ARMY HEADQUARTERS

With Kleist's failure, Oster decided that he had to send another emissary to the British. With only a month remaining before the planned invasion of Czechoslovakia, the conspirators had to know where the British stood. The choice fell on the retired Lieutenant Colonel Hans-Werner Böhm-Tettelbach, an old and trusted friend of Oster's and Beck's with whom they had discussed the possibility as early as August 15.

Böhm-Tettelbach's fluency in English, his military background, and his ties to Beck made him Oster's first choice for a diplomatic mission to London. According to Halder, "Oster sought me out and told me that Beck considered it necessary to send an emissary to London to warn the English against any further giving ground to Hitler." Halder, who had not been informed of Kleist's earlier visit, felt that he had little choice but to approve Beck's plans. He left it to Oster to formulate the instructions to Böhm-Tettelbach.[26]

Oster asked him if he still had contacts in London, and Böhm-Tettelbach replied that he occasionally heard from Julian Piggott, whom he had met when Piggott was British high commissioner of the Inter-Allied Rhineland High Commission at Cologne.[27] Oster told Böhm-Tettelbach to impress on Piggott the need to urge his government contacts to "put a stop to that Hitler policy. If you do, we will act and free you and us from a despot!" Böhm-Tettelbach left for England the next day.

He faithfully discharged his duty, meeting Piggot and an unnamed British intelligence officer, but nothing came of it. He returned to Berlin, and "on a dark night in Elberfelde" gave his report to Oster, who in turn passed it on to Halder.[28]

Piggott and his mysterious guest did not have contacts with sufficient weight in the upper echelons of government to make an impression. With the failure of the missions of Kleist and then Böhm-Tettelbach, Oster must have felt that he needed an emissary who could gain the attention of members of the cabinet. He turned to his newfound allies in the Foreign Ministry, the Kordt brothers.

SEPTEMBER 1, FOREIGN MINISTRY

The historian Carl J. Burckhardt, the League of Nations High Commissioner for Danzig, appeared in Weizsäcker's office on his way to Switzerland. The state secretary seemed discouraged. After only a few perfunctory words at the Foreign Ministry, he asked Burckhardt to meet him half an hour later in the Tiergarten. There, in the vast open spaces of the public park, they could speak undisturbed. Weizsäcker told Burckhardt about the steps that he had already taken to make the British aware of Hitler's intentions. All to no avail. Now, he said, the resisters had a new plan. They would try to reach Halifax and Chamberlain through Theo Kordt, Erich's brother, who was stationed in London. Meanwhile Weizsäcker asked Burckhardt to try to get a message to the British from Switzerland. Tell the British, he implored, to send "an undiplomatic, uninhibited Englishman, a general with a riding crop, for instance," to show up in Hitler's office. Then the führer might listen.

Weizsäcker went on to tell Burckhardt that he was convinced that the only way to save the peace—and to save Germany—would be to eliminate "the one corrupting figure in whom all power was concentrated." In his growing desperation Weizsäcker said that he had been in regular contact with Canaris. The admiral "belonged to the few that I talked to without reservation. Our main topic [was] the prevention of war and the raiding of Hitler's nest."[29]

Weizsäcker also mentioned that he had spoken recently to Beck about his resignation, which by now the general realized had been a mistake. When Weizsäcker had told Beck that he was also thinking of resigning, Beck had talked him out of it.

Through Beck, Weizsäcker was also in close touch with Halder. The two made certain that their correspondence was disguised and that their contact was not obvious, since any communication between the General Staff and the Foreign Ministry was strictly forbidden. Weizsäcker also began regular communications with Canaris, of whom he said: "He was one of the most interesting phenomena of the period, of a type brought to light and perfected under dictatorship, a combination of disinterested idealism and of shrewdness such as is particularly rare in Germany. In Germany one very seldom finds the cleverness of a snake and the purity of a dove combined in one personality."[30]

The contacts among Weizsäcker, Canaris, and Halder cemented the growing relationships among conspirators in the Foreign Ministry, the Abwehr, and the Army General Staff. As the days of September slipped away and the danger of war loomed ever larger, Oster and his collaborators worked with a greater sense of urgency.

EARLY SEPTEMBER, BERLIN, ST. HEDWIG'S CATHEDRAL

One of the most vexing issues confronting German officers was their loyalty oath. Hitler broke his promises as easily as he broke bread, but his officers did not. How could they justify overthrowing a ruler to whom they had taken a personal oath of loyalty?

At Oster's urging Beck arranged for a meeting for several high-ranking officers with the canon of St. Hedwig's Cathedral, Father Bern-hard Lichtenberg, an outspoken dissenter who had openly preached against the Nazis' persecution of the Jews. The five officers—all Protes-tants—heard the Catholic priest's justification for tyrannicide: "A per-son may not be murdered, but a difference exists between such a forbidden murder and self-defense [*Notwehr*]."

There is no record of the officers' reactions, or whether Father Lichtenberg held other such meetings. But it indicates the thorough-ness of Oster's preparations, and his success in bringing together resisters of conscience with military resisters.[31]

SEPTEMBER 3, THE BERGHOF

Hitler had summoned Generals Brauchitsch and Keitel to the Berghof to discuss the invasion plans for Czechoslovakia—"Plan Green." Brauchitsch reported that the assault troops would be moved up to their jumping-off areas on September 28 and simulate maneuvers. From that point on they would be capable of action. When the moment for invasion—X-Day—was ordered, they would simply turn toward the Czech border and launch the attack.

But Hitler had his doubts. The staging areas, he complained, were a good two-day march from the Czech border. Furthermore,

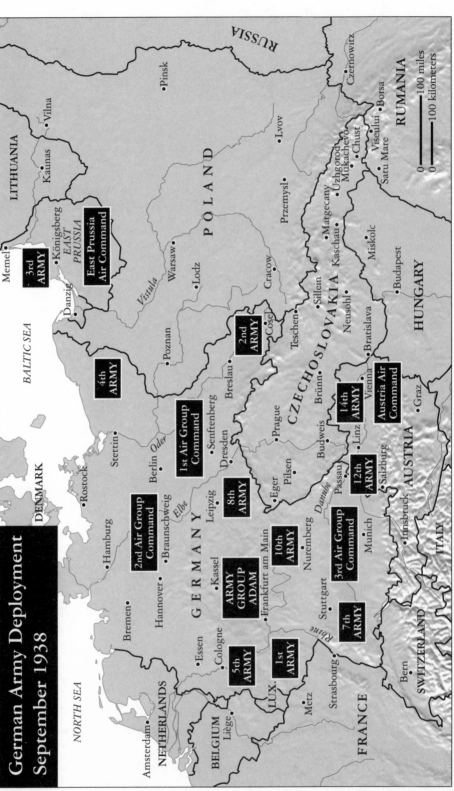

German Army Deployment
September 1938

In dispute between Hitler and the General Staff was whether the *Schwerpunkt* ("point of attack") should come from the 10th Army to the west of Czechoslovakia or the 12th and 14th Armies to the south.
Source: Telford Taylor, *Munich: Price of Peace* (New York: Doubleday, 1979), p. 708.

he didn't like the army's overall plan of attack. Czechoslovakia was a long thin snake, with its Bohemian head and Slovakian tail, hard against Germany's southeastern border. With the addition of Austria, the expanded Third Reich had a more-than-seven-hundred-mile border with Czechoslovakia, offering the attacking Germans many different opportunities. The General Staff's plan was to make the *Schwerpunkt* (point of assault) the Second Army's attack from the north, supported by the Twelfth and Fourteenth Armies' attack from the south, which would form pincers and cut Czechoslovakia in two at its wasp waist. Once Czechoslovakia had been divided, the Wehrmacht could finish off Czechoslovak resistance first in one half of the country and then in the other. Brauchitsch and Keitel were surprised when Hitler grumbled at their presentation, which was based on classic military strategy. It was too predictable, he said. It was just what the Czechs would expect and prepare for, making the Second Army's attack another Verdun, in which it would be bled white.

Indeed, Hitler thought that the main assault ought to come directly from the west, where the Wehrmacht's Tenth Army stood ready at Nuremberg, directly across from the heavily Sudeten German areas on the other side of the Czech border. Transfer the panzer and motorized units to the Tenth, Hitler ordered. The Tenth will break through, and then the southern front will collapse. "A single army in the heart of Bohemia will bring about a decision," Hitler said definitively. Brauchitsch tried to voice his reservations but Hitler merely brushed them aside.[32]

SEPTEMBER 3, BERLIN, ABWEHR HEADQUARTERS

Oster asked Erich Kordt if his brother, Theo, could get a message through to influential British politicians. "If the British government will make a forceful declaration that even a simple man can understand," Oster told him, "you can explain to them that a military faction led by Beck will know how to prevent an outbreak of war. Do you understand?"[33] Erich did.

While the Kordt brothers, along with Weizsäcker, had worked against Hitler's expansionist plans independently from Oster through the summer, by late August they had begun to work together closely. By the time Oster asked him to enlist Theo and his British contacts, Erich was familiar enough with the conspiracy to know that in order for it to succeed, there would have to be careful coordination of foreign action with internal military and political plans.

Kordt meditated on the situation that Hitler had created for him and other like-minded Germans. For the military to overthrow Hitler would take extraordinary courage, but the alternative was to let Hitler lead them into a war that would bring disastrous defeat to Germany, and would probably lead to the gallows for those senior officers who had followed Hitler's orders. "A coup would have been a lifesaver," he concluded. He assured Oster that he would send the message to the British through his brother immediately.[34]

SEPTEMBER 4, BERLIN, HJALMAR SCHACHT'S APARTMENT

Oster set up a meeting between Halder and the most prominent civilian member of the conspiracy, Hjalmar Schacht. With the straightforward style of a career military man, Halder got right to the point. He bluntly asked Schacht "whether he was prepared to take over the administration in case Hitler pursued his course to the point of war and made a violent overthrow of his regime unavoidable."[35] Schacht assured him that he was. Halder was pleased that this internationally known and respected economist would become the face of the new regime.

For his part Oster was elated that Schacht and Halder had cemented their relationship. Halder was the titular leader of the conspiracy. While Oster coordinated the operation from Abwehr Headquarters, he knew that the coup, if it was to have the greatest chance of success with army commanders, would have to be led by someone of Halder's rank and position.[36]

SEPTEMBER 5, BERLIN, HALDER'S APARTMENT

After meeting with Oster and then Schacht, Halder summoned Gisevius to his apartment to discuss police involvement in the coup. To Gisevius's astonishment, Halder himself opened the door. Such was the fear of denunciation in the Nazi police state that Halder did not trust his servants. The general spoke with Gisevius for several hours "with singular frankness."[37]

Gisevius noted that despite Halder's small stature and "stolid" appearance, when he spoke about Nazi injustice, "suddenly he was all fire and fury." Halder called Hitler "this madman, this criminal" who was taking Germany into war "possibly because of his 'sexually pathological condition' which created in him the desire to see blood flow." Like Oster, Halder cited the events of the Night of the Long Knives and the "countless murders in the concentration camps" as evidence of Hitler's criminal derangement.

Halder said that while he had no illusions about Hitler's intentions, many other officers did. They were misled by the saying, then circulating among the officer corps, that Hitler's preparation for war "was all just a colossal bluff," and that Hitler was "merely preparing a diplomatic maneuver of blackmail on a large scale."

Halder and Gisevius began by discussing the possibility of an army attack on SS and Gestapo Headquarters, which would serve to rally the troops against an undeniable foe but would leave Hitler untouched. While neither man specifically addressed the issue of how a coup could succeed while Hitler was still alive and at large, it hovered over the remainder of the conversation like an unacknowledged ghost.

Gisevius argued that a coup aimed at the SS and Gestapo could be launched immediately, because the behavior of Himmler, Heydrich, and their minions was so ignominious that the army would rally to it without hesitation. Halder disagreed, arguing that Hitler's popularity was so great that "it would take a succession of bad experiences to enlighten the army. What was needed," he proposed, "was a setback, such as no propaganda tricks could dissimulate, to induce the army to cooperate with an uprising against Hitler."

Halder saw the best chance for a setback in foreign affairs. In particular he expected that the Western Powers would intervene and declare war against Hitler if he invaded Czechoslovakia. Only at that point, Halder argued, could the conspirators take action. Halder understood that this was risky because the invasion might actually begin before a coup could be launched. But he thought that such a risk was worth taking because "until the people were given tangible evidence that Hitler wanted war in bloody earnest, they would not snap out of their state of permanent intoxication."

They also discussed how they would deal with Hitler in the event of a declaration of war and a coup. Halder wanted Hitler to meet with a "fatal accident." He thought that Hitler's train should be blown up and the rumor spread that the führer had been killed by an enemy air attack. Gisevius was disconcerted by Halder's infatuation with this indirect and uncertain means of assassination, particularly since Halder carried a pistol and had frequent access to Hitler, and could kill him at one of their meetings.

The next morning Gisevius got in touch with Oster. Oster informed Gisevius that Halder had been so impressed with him that he had ordered the two to prepare "an outline of all police measures to be taken in the event of a *coup d'état*." By early September, with an energetic Halder making arrangements with other key members of the conspiracy, Oster and Gisevius began to think that the much-discussed coup might actually take place.

September 7, London, Number 10 Downing Street

When Theo Kordt entered Sir Horace Wilson's office on the morning of September 7, Halifax was already there. Kordt knew that this would be his best—and perhaps his last—chance to make the British understand the extreme urgency of taking a stand against Hitler before he launched the invasion of Czechoslovakia.

The German diplomat told Halifax that he came to him "as the representative of political and military circles in Berlin who wanted to prevent war by all means." Kordt informed Halifax that Hitler intended to attack Czechoslovakia within the month, and "the political and military circles I am speaking for strongly object to that policy. We think that giving way to Hitler's policy of force in this moment would definitely mean barring the way for the return to the conceptions of honour and integrity amongst European nations."[38]

Kordt reminded Halifax that Europe had slipped into war in 1914 in part because British Foreign Secretary Sir Edward Grey had not definitively warned the German government that in case of a Franco-German war, Great Britain would assist France. Perhaps if Kaiser Wilhelm had understood this, Kordt speculated, he would not have pursued war so recklessly. To avoid a replay of this historical tragedy, Kordt urged the British government to warn Hitler that an invasion of Czechoslovakia would mean French *and* British intervention—in essence the beginning of another world war.

"Should, however, Hitler persist in his bellicose policy," Kordt continued ominously, "I am in a position to assure you that the political and military circles I am speaking for will 'take arms against a sea of troubles and by opposing end them.'" Lest Halifax fail to grasp his Shakespearean allusion, Kordt summarized his conclusion bluntly: "In German public opinion as well as in responsible army circles Hitler's war is unpopular and considered a crime against civilisation. If a statement as required were issued [by the British government] the army leaders are prepared to act against Hitler's policy." Kordt thus put the position of the conspirators even more candidly than had the previous emissaries.

Halifax listened to Kordt's impassioned plea "with sincerity and excitement," according to Theo. "He thanked me for the courage with which I had spoken at this highly critical moment." Halifax's most promising statement to Theo, however, was his blunt statement that "war would become impossible to prevent if Hitler used violence against the Czechs," a much stronger position than His Majesty's Government had previously taken. At the end of the conversation Halifax promised to convey Theo's message to the prime minister and to several of his cabinet colleagues.

Following the meeting Theo Kordt was optimistic. He had delivered the conspirators' message to Lord Halifax, who had been impassive—his usual state—but interested and reasonably supportive. Based on Halifax's statement about the inevitability of war if Germany tried to solve the Sudeten issue by force, Kordt believed that His Majesty's Government would issue the clear warning to Hitler that he had urged on them. Later he would recount, somewhat wistfully, that "those of my friends who lived in the political atmosphere of London during the critical September days [of] 1938 quite shared my opinion that the British government had really made a clear stand." With that understanding, Kordt hurried off to rendezvous with his cousin Susanne.[39]

SEPTEMBER 7, LONDON, THE TRAVELLERS' CLUB, PALL MALL

The Travellers' Club was one of the lairs of privilege in central London where wealthy and powerful Englishmen gathered to sort out the affairs of state. On this day, two members of the Tory establishment were having lunch. Geoffrey Dawson, editor in chief of *The Times*, had momentarily become the most controversial man in Britain. His editorial that morning had recommended that Czechoslovakia might be made "a more homogeneous state by the cession of that fringe of alien populations who are contiguous to the nation to which they are united by race." In other words Dawson was advocating that Czechoslovakia cede the Sudetenland to Germany and other pieces of its territory to Hungary and Poland.

Across the table, fresh from his secret meeting with Theo Kordt, was Lord Halifax. In contrast to the professional diplomats of the Foreign Office, he did not seem particularly bothered by Dawson's proposal to dismember Czechoslovakia. At least that was Dawson's perception. "The Foreign Office went through the roof," he remembered. "Not so, however, the Foreign Secretary, who came and lunched with me at the Travellers', and had a long talk. He is as much in the dark as everyone else as to what is likely to happen next."[40]

If Halifax was calm in the face of Dawson's editorial, the European diplomatic community was not. Diplomats in Prague and Paris were agitated, and Theo Kordt suspected that the editorial had been inspired by "the Prime Minister's entourage." By reiterating the call for cession in *The Times* the next day, Dawson did nothing to allay the concern that he was speaking for His Majesty's Government.

SEPTEMBER 8, BERLIN, III ARMY CORPS HEADQUARTERS

The conspirators focused their hopes on General of Infantry Erwin von Witzleben, commander of the Berlin Wehrkreis. His troops would have to seize and secure key points in and around the capital. Gisevius said of him, "Witzleben was a refreshingly uncomplicated man. . . . The Berlin commander was a typical front-line general who had his heart in the right place. Probably not too well read and certainly not inclined toward the fine arts, he was nevertheless a man firmly rooted in the chivalric traditions of the old Prussian officers' corps."[41]

Oster had served under Witzleben's command in the 1920s and thought that, had he not been seriously ill during the Fritsch crisis, Witzleben would have led a revolt against the regime. Now Oster told him about the conspiracy.[42]

Witzleben asked Oster if there would actually be war, or if the diplomatic disturbances were just Goebbels's usual "stage thunder." He also asked about Hitler's oft-repeated hints to army officers that there was a secret agreement between him and the leaders of the Western Powers "under which Germany was to defend Europe against Bolshevism." If such an agreement existed, it would explain British and French inaction in the face of Hitler's aggression, which otherwise seemed incomprehensible to Witzleben and other generals. Oster assured the commander that there was no secret agreement. Hitler's hints were simply another way that he had misled his generals. At the end of the conversation, Gisevius reported, Witzleben "placed himself unconditionally at our disposal."[43]

SEPTEMBER 8, NUREMBERG

The annual Nazi Party rally had been proceeding at full force since September 5 in Nuremberg. On the vast parade grounds on the outskirts of the ancient city, where Leni Riefenstahl had filmed *Triumph of the Will* just four years earlier, hundreds of thousands of Nazis marched, saluted, and chanted by torchlight. The world waited nervously, expecting Hitler to use his speech on September 12 to declare war on Czechoslovakia.

General of Artillery Alfred Jodl had accompanied Hitler and his entourage to the rally, but the general had more on his mind than the celebration of Nazism. Planning for the invasion of Czechoslovakia continued at an intense pace and he was at the center of it.

Jodl was visited by Lieutenant General Carl-Heinrich von Stülpnagel, Halder's deputy chief of the General Staff, who came to request that the army be given at least five days' notice before the invasion would begin. Jodl agreed, although he added the proviso that if it were impossible to predict weather conditions for more than two days in advance then that would be the most warning that he could give the General Staff.

Stülpnagel, who was also in on planning the conspiracy, undoubtedly was relieved to be able to assure Halder that he would probably have five days', but at least two days', notice before X-Day. Nevertheless, Stülpnagel expressed his doubts to Jodl about whether the invasion plan could succeed, particularly since it depended upon the dubious assumption that the Western Powers would not intervene. Jodl told Stülpnagel to keep a stiff upper lip. But to his diary he confided his own doubts: "I must admit that I too am worried, when comparing the changing assessments of our political and military potential as between the [earlier] directives and . . . the latest pronouncements."[44] Doubts were creeping into the minds of even the führer's closest military associates.

SEPTEMBER 8, NUREMBERG

General Curt Liebmann had been ordered to attend the Nazi Party rally, as he had every year since 1933. He ran into Halder, and "[their] conversation understandably turned to the Czech question."[45] Halder shared his concern about the weakness of the West Wall, the lack of sufficient matériel, and the army's general lack of readiness. When Liebmann asked Halder how it was that Brauchitsch could have failed to convey this view to Hitler, Halder answered, "There is only one way to stop the course of things, and that would be the removal of Hitler by forceful means." Liebmann said that the timid Brauchitsch was certainly not the man for such a task, and Halder answered, "No, but we have a qualified person who would be ready to do it." Was it Fritsch, Liebmann asked? "Halder made a confirming gesture but added that he could say no more."[46]

SEPTEMBER 9–10, LONDON/NUREMBERG

Halifax was in a stew about whether to send a clear warning message to Hitler. In his communications with Chamberlain, Churchill, and Theo Kordt over the past ten days, that had been the central question. Just that afternoon Anthony Eden had called on him at the Foreign Office, and had warned him that, in Eden's opinion, it was "inconceivable" that a war between France and Germany over Czechoslovakia could be localized. Inevitably Britain would be involved "up to the neck." According to Eden, Halifax had replied, "Great minds are thinking alike, for my mind is moving on just such a project and indeed I was going to speak to Neville about a draft today."[47] He asked Eden for notes of his proposals.

Despite his disavowal of such action at the cabinet meeting of August 30, Halifax had recently been entertaining the idea of giving an unequivocal warning to Hitler in a public speech, as Kordt had urged him to do. Halifax had actually drafted such a speech, which had frightened both Simon and Chamberlain, who had forced him to give up the idea. As an alternative Halifax considered a private warning to Hitler. If he sent it now, before Hitler spoke, perhaps he could dissuade the führer from making an ultimatum that would lead to war. However, as Halifax knew, sending a warning was dangerous. It would virtually commit Britain to go to war if Hitler disregarded it and attacked the Czechs.

Finally, on September 9, against Chamberlain's counsel but with his grudging consent, Halifax drafted a warning message to Hitler, and sent it by courier to Ambassador Henderson, who was staying at Nuremberg in the uncomfortable confines of a train carriage. (All hotel rooms had been booked by the party faithful.) Henderson read it and was appalled. He had recently delivered warning messages to Ribbentrop and Weizsäcker, but he categorically refused to deliver this one. The ambassador claimed that "the form of Hitler's genius is on the borderline of madness. He may already have stepped over the edge as some people believe. . . . A second 21st May [that is, a strong warning against invasion] will push him over the edge. That I truly and honestly believe: if not of actual madness, of mad action."[48]

It was breathtaking: The ambassador of His Majesty's Government refused to deliver a diplomatic note because he feared that its intended recipient might thereby be driven into full-fledged madness! British foreign policy was being dictated by Henderson's psychiatric evaluation of Hitler's personality.

Chamberlain, when he heard of Henderson's reaction, withdrew his earlier consent and insisted that Halifax take back the note. Halifax's resolve melted. Faced with a decision that might plunge Britain into war, Henderson was certain; Halifax and Chamberlain were not. The insubordinate ambassador prevailed over the indecisive ministers.[49]

SEPTEMBER 9–10, NUREMBERG, 10 P.M.–3:30 A.M.

Just as British politicians were dithering about whether or not to send a message that might disturb the führer's precarious mental state, the führer himself was planning war at one of his night-owl sessions. Brauchitsch and Halder—the latter on his maiden voyage as chief of the General Staff—marched grimly into Hitler's headquarters. Halder had been infuriated by Hitler's cavalier change of the General Staff's invasion plans six days earlier.

Halder now formally presented the General Staff's version of Plan Green, emphasizing once again the primacy of the pincer movement by which the German Second Army, attacking from the north, and the Fourteenth Army, attacking from the south, would cut Czechoslovakia in two. He said that Hitler's plan to make the Tenth Army the *Schwerpunkt* from the west was flawed. The attacking Tenth Army would face strong fortifications and be hindered by bad roads.

Hitler heard Halder out but was unconvinced. He noted the strength of the Czech fortifications facing the Second Army, and once again he raised the specter of Verdun and used the words that brought back haunting memories of that murderous battle to all German soldiers: "bleeding to death." He lectured the generals—as if addressing two dull-witted schoolboys—about the wider implications of the campaign. Their plan was too uncertain and too likely to get bogged down; for political reasons he must have a decisive victory within eight days.

In the end Hitler rescinded some of his earlier orders about the transfer of armored and motorized units, but he continued to insist that the Tenth's attack from the west must be carried out *in addition to* the attack from the north and south. Then, just to give his generals a sense of how he valued their professional expertise, he lectured them about cavalry tactics during 1914 and the lessons to be learned there about contemporary motorized unit tactics.[50] Halder returned to his quarters more convinced than ever that this unthinking warmonger would lead Germany to its doom.

SEPTEMBER 10, BERLIN, POLICE PRESIDIUM, ALEXANDERPLATZ

Gisevius at last secured Oster's permission to contact Count Helldorf, the president of the Berlin Police. Helldorf was a rough-hewn individual who was also an SA *alter Kämpfer* (old fighter). Although both Gisevius and Oster understood the importance of securing police cooperation, they agreed that an approach to Helldorf must be made with extreme caution. Gisevius noted that "after years of feeling [him] out, I at last talked plainly with Count Helldorf." Much to his delight, Helldorf immediately "declared himself willing to go along with us."[51] Still, Gisevius was cautious and did not lay all of his cards on the table. He did not tell Helldorf, for example, about Witzleben's role in the impending coup. Meanwhile, Fritz von der Schulenburg, the vice president of the Berlin police, had already joined the conspiracy, and throughout September played an important role as messenger between Kordt and Oster.

Gisevius's and Oster's success in securing the active cooperation, or at least the neutrality, of the Berlin regular police (Kripo)—as opposed to political police (Gestapo)—was crucial to the success of the plan. It meant that General Walter von Brockdorff-Ahlefeldt's troops could concentrate their efforts on SS stations and other targets, knowing that the police would not interfere with them. Indeed, Gisevius called Schulenburg's and Helldorf's cooperation "the greatest moment for our action in Berlin."

SEPTEMBER 10, NUREMBERG

Erich Kordt had been at the Nazi Party rally since its opening. On September 9 he had learned that Susanne Simonis had returned to Berlin, and that his brother, Theo, had delivered the message to Lord Halifax on September 7. But in the intervening forty-eight hours there had been no definitive message from the British.

Would the British respond to the conspirators' overture? That evening Kordt met with his friend Philip Conwell-Evans, who was at the party rally, probably at the request of Vansittart. "We agreed that he would travel to London the next morning," Kordt remembered, "and try to get through to the prime minister. If the British government wanted to keep the peace it should speak immediately."

Kordt's spirits were momentarily lifted when he heard that a message had been received from London, but on closer inspection it was not a statement from Halifax, but merely a press release about Czechoslovakia from the Foreign Office that was so finely nuanced as to be meaningless. Dictators, Kordt thought, understood only the language of force, not diplomacy.[52]

SECOND WEEK OF SEPTEMBER, BERLIN

During the second week of September, according to Gisevius, with the invasion due to begin within weeks, the detailed planning of the logistics for the coup began in earnest.[53] Once Oster had secured General Witzleben's assurance that he would participate, he arranged for Gisevius to bring together the two most prominent members of the conspiracy, Schacht and Witzleben.

The crucial meeting was held around September 10, "on a quiet afternoon at Schacht's country home northeast of Berlin," with Gisevius and Generals Witzleben and Brockdorff attending. Witzleben charged Brockdorff, whose crack Twenty-third Infantry Division was considered politically and militarily reliable, with seizing key objectives in and around Berlin.[54] Gisevius was impressed by Witzleben's resolve. "With or without Halder, on orders from above or against the orders of his own military superiors—he was ready to go the limit."[55]

Gisevius and Brockdorff-Ahlefeldt were to draw up detailed plans, the latter for military action and the former for police action. Schacht would make a list of political administrators who would be called on once the coup had succeeded, "but this last seemed to us to be the least of our worries." All the participants assumed that once the army had launched the coup there would be a "state of siege" that would last for a few days, but after that power would be returned to civilians.[56]

Erich Kordt was certain that a provisional government would consist of persons who were different from the conspirators. He thought that it would take between two weeks and two months for such a government to form. In the end "the political forces had to rally anew, which could be done only after the downfall of the regime," which was his only interest.[57]

The other conspirators did not trouble themselves about the political details, according to Gisevius. "What persons were to compose the new cabinet would be settled after Hitler and the Nazis were safely out of the way."[58] On this issue the men of 1938 avoided the trap of later conspirators, who debated endlessly and fruitlessly the hypothetical composition of postcoup governments. Gisevius noted laconically, "There was still a sufficient reserve of well-known statesmen and officials to fall back on."[59]

A more divisive issue was how to deal with Hitler. As most German people still revered Hitler and were unwilling to believe that the führer wanted war, one possibility was to direct the coup at Himmler and Heydrich rather than at Hitler. "People believed Himmler capable of any crime, including an attack upon Hitler or the Army leadership. Invoking the Army to protect its supreme commander against an SS plot was therefore not so fantastic. In fact, it was credible."[60] Gisevius thought that at the onset of a coup, all SS and Gestapo men should be ordered to report to local army headquarters. He believed that "unquestionably the majority would have reported at once, with a thousand assurances of loyalty," certainly an optimistic prediction.[61]

But a coup ostensibly made against the SS still did not settle the awkward question of what to do about Hitler. Gisevius said that his views on this subject differed from his friends'. While some conspirators wanted to take Hitler alive and try him in court, and others wanted to declare him insane, Gisevius favored assassination. "I held that tyrannicide had always been looked upon as a moral commandment."[62] A führer who was still alive, even in jail, was a danger. "There was far too much at stake in a *coup d'état*; rebels could not afford the moral luxury of giving their main enemy even the smallest opening by which to escape and possibly launch a counterattack."[63]

Witzleben gave Gisevius an office adjacent to his own at military headquarters to continue his planning, and told his adjutant with a wink that Gisevius was a relative who was filing the family papers. Sequestered in this safe haven, far from the eyes and ears of the Gestapo, Gisevius worked on classified information that Oster and Nebe had procured, detailing the strength and location of SS formations throughout the country.

Although Gisevius saw several problems connected with the seizure of power, he believed that once it had been accomplished, the government would function without serious interruption. Nazi control of government agencies was superficial: "It would be necessary to replace a few chief men in the various ministries and the apparatus would continue to run smoothly on the momentum characteristic of all bureaucracies."[64]

A more difficult issue was military action. Gisevius assumed that the "strike force" would be limited to no more than two army divisions. The conspirators could expect no civilian assistance. By 1938 potential rebels—such as trade unionists or anti-Nazi politicians—had all been crushed. "Any general who chose to act was pretty much thrown on his own resources. At most he could hope that the spark would leap to some of his fellow generals—and that depended largely upon the success he achieved during the first few hours of his rebellion."[65]

Within twelve hours, Gisevius thought, the rebel leaders could expect to face Nazi opposition, perhaps "even before the two divisions could unite their forces." This scenario ruled out a coup that was launched from the provinces. The first blow *must* be struck from the capital. According to the plan Witzleben's troops would occupy the headquarters of the federal police. Once that had been accomplished, they could storm the Reich Chancellery and other government offices. Surprise was of paramount importance, Gisevius thought; otherwise "the most important birds would already have flown."[66]

The conspirators believed that while the seizure of power in Berlin was in progress, it was essential that generals in the provinces join the revolt or at least remain neutral. Such was the level of dissatisfaction that had been building among local Wehrkreis commanders, coupled with the recruiting efforts of Oster and his closest colleagues, that Gisevius estimated there was "a reasonable prospect of success" of securing the support of those provincial generals who actually commanded troops.[67] Of course Oster and Gisevius knew that Lieutenant General Erich Hoepner's First Light Division was ready to join the action if the only SS regiment, the Leibstandarte Adolf Hitler, tried to race back to Berlin. Furthermore the conspirators also had confirmation that Major General Paul von Hase, commander of the Fiftieth Infantry Regiment at nearby Landsberg an der Warthe, was ready to spring into action.[68]

By early September, despite Beck's resignation, Halder's enthusiastic participation connected the Oster circle to the upper echelons of the army hierarchy. Witzleben and Brockdorff-Ahlefeldt were committed to leading the coup in Berlin. Most important, detailed planning about the logistics of the coup was finally under way. Oster had good reason to feel optimistic.

SEPTEMBER 11, BERLIN, ARMY HEADQUARTERS

To celebrate the birth of their first grandchild, General Erwin von Witzleben and his wife were entertaining his cousin, Lieutenant Colonel Hermann von Witzleben, and his wife, Ursula, at their apartment. Ursula remembered the day as very cold and windy. Thus she was surprised when her cousin asked her to take a walk in the garden with him. He said that he wanted to speak to her confidentially, and he thought that his apartment was probably "full of microphones."[69]

Erwin told Ursula bluntly, "Hitler has to go." He said that one solution would be for him to confront Hitler "man-to-man" and tell him to resign. Ursula replied, "A responsible general is no longer going to the King of Prussia but to Adolf Hitler. You will go into the Reich Chancellery and never come out." The general admitted "that is what the others [presumably his fellow conspirators] said. The only thing left is assassination." Ursula was shocked by this idea and asked if there wasn't any other way. No, the general said, "We have examined our consciences. There is no other way. Hitler wants war, he is provoking it, [and] it will be the end of Germany." He added prophetically, "People will have to carry the burden of collective guilt, but they will not understand it and will not be able to carry it."

Then Erwin got down to the business of the conversation. He needed to know which of the leading generals would support a putsch. In particular, he wanted Ursula's opinion of Colonel General Rundstedt. Ursula had been reared in Rundstedt's home and understood his character. She replied that while he had many favorable qualities, he was a vain and ambitious man. "I could only advise my cousin against informing him of the plans." She was quite certain that Rundstedt would not betray the plotters, but "he would never participate." Erwin also asked her about Generals Fromm and Fritsch, the latter so recently humiliated by Hitler. While she disliked and distrusted the former, she had very positive feelings toward Fritsch. In fact, she was left with the impression that Witzleben had already informed Fritsch about the developing plot.

The conversation ended by Erwin asking her not to tell her husband, Hermann, "because if the plan does not succeed, the first thing that would happen is that the second highest ranking officer [in the family] would be arrested and it is my desire that if I am out of commission, that he will carry on with what we want." Ursula agreed and, somewhat shaken by these revelations, accompanied the general back to the party.

SEPTEMBER 11, BERLIN

In a cloak-and-dagger tactic out of the Abwehr's playbook, Oster recruited Elisabeth Gärtner-Strünck to drive Brockdorff and Gisevius through the streets of Berlin and its environs on a reconnaissance mission.[70] The Strüncks were family friends of the Osters—they had skied together in the Dolomites the previous winter—and were comparatively wealthy and strongly anti-Nazi. They moved to Berlin specifically to lend assistance to the conspirators. While a general riding around Berlin in an army vehicle reconnoitering government buildings would immediately arouse suspicion, three civilians in a private car would pass unnoticed.

According to Gisevius, he and General Brockdorff met Frau Strünck at a suburban train station and "she drove the two of us, harmless sightseers that we were, through the German capital in her handsome automobile." Brockdorff asked her to drive several times around each of the potential target buildings "in order to spot possible escapes through gardens and neighboring structures."[71]

Frau Strünck remembered that their extended ride took them to the government buildings on Wilhelmstrasse, Göring's palace, the Lichterfelde SS Barracks, Sachsenhausen concentration camp, and the Königs Wusterhausen radio station, locations widely scattered throughout Greater Berlin. As they homed in on their targets, they realized that the decision to strike simultaneously "would require an unusually large number of soldiers." Finally, Gisevius and Brockdorff realized how important it was to have secured the cooperation of the Berlin police. "Without the police we should have to divide our army forces into too many small units."

In the wake of the clandestine automobile ride, Brockdorff quickly finished his work. Several days later Witzleben announced that

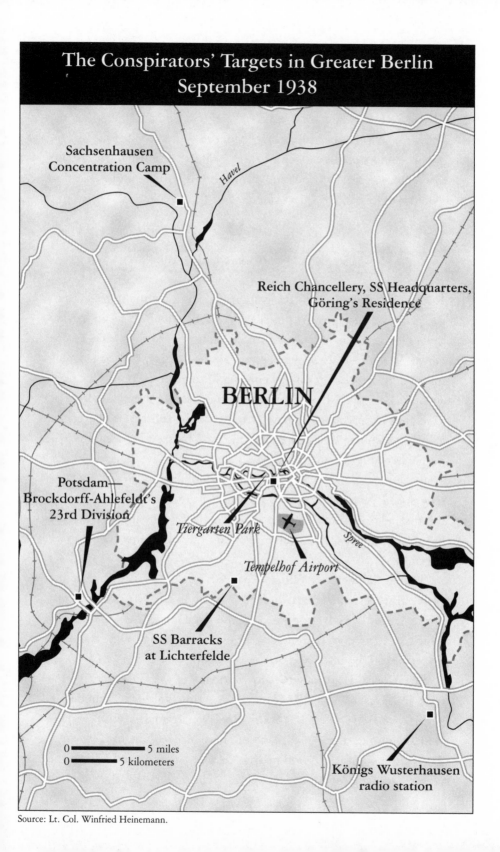

The Conspirators' Targets in Greater Berlin
September 1938

Sachsenhausen
Concentration Camp

Havel

Reich Chancellery, SS Headquarters,
Göring's Residence

BERLIN

Potsdam—
Brockdorff-Ahlefeldt's
23rd Division

Tiergarten Park

Spree

Tempelhof Airport

SS Barracks
at Lichterfelde

0 ————— 5 miles
0 ————— 5 kilometers

Königs Wusterhausen
radio station

Source: Lt. Col. Winfried Heinemann.

the military plans for the coup were complete; "their realization needed only the push of a button."[72] Erich Kordt claimed that the "provisional date" for a revolt was "between September 14 and 16," no doubt a reflection of the conspirators' belief that Hitler might use his forthcoming speech to deliver an ultimatum to Czechoslovakia that would lead to war a few days later.[73]

SEPTEMBER 12, NUREMBERG, ZEPPELIN FIELD

The führer strode through the colonnade and down the stone steps to the platform jutting out from the spectators' tribune. He nodded to those closest to the aisles as he passed, and received the cheers of the crowd. When he reached the podium he stood silently and waited, illuminated by the two enormous burning braziers on pylons to his right and left. Thousands of SS men stood at rigid attention, and hundreds of thousands of spectators waited excitedly for the führer's words. Hitler had always known how to play a crowd, and his speeches were theatrical performances created by a master dramaturge. Every hand gesture, every hesitation, and every "spontaneous outburst" was carefully choreographed. As he had done during his drive toward power, he would use the radio to reach an even larger audience. But those who only heard and did not see the führer in action missed an important part of the show.

Quite deliberately Hitler began by speaking softly, so that the crowd had to strain to hear him. "So today in the National Socialist Reich we see ourselves opposed by the same forces, the same factors which, as a Party, we had the opportunity of coming to know during fifteen years," he intoned. "So far as this does but give a general witness to the hostile attitude towards Germany of the democratic countries it leaves us cold."[74] Then, with rising voice, he thundered, "Today we are insulted, but thank God! We are in a position to prevent any plundering of Germany or any violence done to Germany." The führer went on to outline the "horrors" perpetrated against the ethnic Germans of Czechoslovakia. "The misery of the Sudeten

Germans is indescribable. It [the Czech state] has sought to annihilate them. As human beings they are oppressed and scandalously treated in an intolerable fashion."

Hitler saved much of his venom for "those other democracies"— Britain and France—who supported the Czechs. "I can only say to the representatives of these democracies that this [the condition of the Sudeten Germans] does concern us, and if these tortured creatures can find no justice and no help they will get both from us." Later in the speech he repeated this point, allowing his voice to rise to a nearly hysterical pitch. "If . . . they [the British and French] must support [by] every means the oppression of the [Sudeten] Germans, then this decision will have serious consequences!" He added ominously, "I serve peace best if I leave no doubt on this point." The crowd roared as Hitler yelled, "The German Reich has slumbered long. The German people is now awakened and has offered itself as a wearer of its own millennial crown."

In the wake of the führer's incendiary speech, riots erupted throughout the Sudetenland, and the Czech government declared martial law, creating an obvious excuse for Hitler to intervene to save the Sudeten Germans from Czech "oppression."

SEPTEMBER 13, BERLIN, OKW HEADQUARTERS

Jodl confided to his diary that while Hitler's speech had been "magnificent revenge" on the Czechs, many among the officer corps should "blush with shame for their faintheartedness." Keitel had also noted the "killjoy" spirit of many of the generals. Jodl gloomily concluded that "the atmosphere in Nuremberg is cold, and the führer has the whole nation behind him but not the leading generals of the Army. The generals do not accept the 'genius of the führer' and thus do not obey him."[75] With war looming, even Hitler's closest military associates worried about the loyalty of their colleagues.

SEPTEMBER 13, BERLIN, REICH CHANCELLERY

Paul Schmidt, Hitler's interpreter and one of the conspirators' informants, returned to Berlin "with Hitler's threats still ringing in my ears." He found that his closest friends and colleagues were gripped by "profound depression," believing that war was inevitable. At this time he "also learnt of the opposition's plan to have [Hitler] summarily arrested by the army as soon as he ordered general mobilisation." Even Schmidt, on the periphery of the conspiracy, knew that action against the regime was likely at any moment. By now, he commented, "the tension had become unbearable."[76]

SEPTEMBER 13, BERLIN, HALDER'S APARTMENT

With the coup now perhaps imminent, Oster wanted to make certain that all the principals were agreed on the details. Since Gisevius and Schacht felt that their separate interviews with Halder earlier in the month had left some issues unresolved, Oster set up an appointment for them to pay a return visit to the chief of the General Staff. Halder remembered that Schacht brought along Gisevius without previously informing him, "much to my annoyance."[77] Perhaps this unwelcome surprise, combined with the tension that had prevailed at the party rally for the previous few days, unsettled the often mercurial Halder. "The interview was a stormy one," Gisevius noted.[78] Both Schacht and Gisevius formed the impression that Halder had an intimation that Britain might "present Hitler with a free ticket to the East," which would undercut the conspirators' plans. This led Halder to launch into "violent and unexpected denunciations of British policy."

After enduring Halder's outbursts, they got down to specific issues. The main problem was about what event should trigger the coup d'état. "The great difference between us was that we were unwilling to make the outbreak of war an essential condition of our preparations." Gisevius and Schacht did not want to rush into a coup, but they also did not want it to depend on an event so fraught with uncertainty as the outbreak of a war.

Halder heard them out but was firm. According to Gisevius, the general assured them that "the coup d'état was now scheduled to take place in the brief breathing spell between Hitler's final order for the troops to march and the first exchange of shots." Gisevius noted that the risk in this plan was that it "depended upon the dictator's not taking us by surprise with his *Blitz*." Gisevius maintained that Halder said that was impossible. " 'There's no chance of his tricking me. I've arranged the [invasion] plans in such a way that I cannot help having three days' warning before any action is taken; moreover, the final order must be issued directly to me at least twenty-four hours beforehand.' "

Halder also assured them that he had taken steps to counter the only significant SS threat. The First Light Division, under Lieutenant General Erich Hoepner, had been ordered to Thuringia, south of Berlin. While the formation was ostensibly on maneuvers, it was actually positioned to block the only heavily armed SS formation, Leibstandarte Adolf Hitler, stationed near the Czech border at Grafenwöhr training grounds, should it try to reach Berlin when the coup erupted.[79]

Lurking just beneath the surface of this tense conversation, according to Gisevius, was the "painful question of whether Halder was really in earnest," a doubt shared by Oster. In the excitement at the beginning of a war, would Halder cite "strategic considerations" as a reason for deferring the coup? And would the expressed support of other generals also melt away as the reality of preparing for war took precedence over their political concerns? Halder's caution caused Oster, Gisevius, and Schacht to suspect that "this uniformed dialectician was unreliable."[80]

SEPTEMBER 13, BERLIN, ABWEHR HEADQUARTERS/FOREIGN MINISTRY

Erich Kordt returned from Nuremberg to Berlin and tried to find out what reaction Hitler's speech had elicited from his military colleagues. He

went to Abwehr Headquarters seeking Canaris. Instead Kordt encountered one of Oster's friends, who told Kordt that Oster was expecting that
the coup would be carried out shortly, particularly since Hitler had made
a bad impression on the generals with his Nuremberg speech.

Kordt then returned to the Foreign Ministry, where he met with
Weizsäcker. His superior said that Halder had told him that Hitler
would be arrested at any moment. "The plan had envisaged that Hitler,
while under arrest, would give consent to a change of regime—which
would thus have been brought about more or less legally." After the
seizure of power had taken place, Weizsäcker noted, Hitler's fate "could
be decided upon at the time."[81]

After the tense and confusing days at Nuremberg, Kordt was
becoming more confident that the coup would actually be launched.

SEPTEMBER 14, BERLIN

As the Sudetenland boiled over with clashes between rioting Sudeten
Germans and Czech policemen, Hitler finally decided to issue mobilization orders, which the conspirators learned about immediately.
This ratcheting up of war preparations created panic throughout the
Third Reich. According to Gisevius, "Even the Nazis were frightened out of their wits." The conspirators' politically quiescent friends
"ran about imploring the general[s] to come out and save them."
Gisevius and the rest of the Oster circle listened carefully but said
nothing for fear of giving away information about the plot. "In those
days we pretended to be the loyalest of the loyal," Gisevius noted
wryly.[82]

SEPTEMBER 14, LONDON, TRAVELLERS' CLUB

Theo Kordt met Sir Horace Wilson for lunch. They had been meeting
periodically since the end of July, but deteriorating Anglo-German
relations had led to a fraying of nerves. Sir Horace whined about the

Nazis, "We treat them like gentlemen and they are gangsters." Kordt snapped back, "And when there were gentlemen in the German government, the British government treated them like gangsters."[83]

September 14, Berlin

Liuetenant Colonel Helmuth Groscurth, an Abwehr officer, asked his brother and sister-in-law, "Can you keep a secret? Hitler will be arrested tomorrow."[84] This indiscretion indicates the dilemma that Oster faced in September. In order to gather support for the imminent putsch, he had to widen the network of people who knew about the plans. But that created increasing dangers as even men of goodwill, like Groscurth, could not help blurting out what they knew to their friends and family members.

September 14, London, Number 10 Downing Street

The ministers of His Majesty's Government, like the rest of the world's politicians, had listened nervously to the translation of Hitler's speech two nights earlier from the Nazi Party rally. They had heard the führer denounce the Czechs and hint darkly about the possibility of German intervention. But Hitler had not declared war. Given their fears and expectations, cabinet members counted it a victory for moderation.

In the meantime a new idea had been circulating about how to deal with the Sudeten Germans. At the outset of the Runciman mission, nearly two months earlier, most people had expected Runciman to broker a deal in which the Sudeten Germans would be given greater autonomy within the Czech state. Not even Henlein and the Sudeten leaders, in their radical "Karlsbad demands," had dared to ask for secession from Czechoslovakia. Now, after Dawson's call for secession in *The Times* on September 7, the idea was in the air, under the cover of a "plebiscite" on whether the inhabitants of the area wished to live under one regime or another. (Plebiscites had been used in the Saar, to return that territory from French to German administration in 1935, and in Austria to ratify the Nazi takeover in 1938. Although the Nazis had used them for their own narrow interests, plebiscites had credibility among Europeans who still embraced President Woodrow Wilson's principle of "national self-determination.")

The cabinet meeting opened at 11 A.M. Chamberlain noted that while Hitler had not specifically committed himself to a plebiscite in his speech of September 12, "he used words which pointed in that direction." Chamberlain conceded that while the practical problems of holding a plebiscite were significant, nevertheless "it would be difficult for the democratic countries to go to war to prevent the Sudeten Germans from saying what form of Government they wanted to have."[85] In this offhand way the prime minister announced that he was ready to concede a plebiscite that would lead to the transfer of the Sudetenland to Germany.

Then Chamberlain warmed to his real topic: the unveiling of "Plan Z." He noted that Ambassador Henderson had opined that a surprise visit from the prime minister on the eve of a German invasion of Czechoslovakia "might cause [Hitler] to cancel that intention." Although Henderson played into Chamberlain's White Knight fantasy, he at least had the good sense to warn him about the unpleasant possible outcomes of an unannounced visit. Chamberlain reluctantly agreed. Chamberlain apologized to cabinet members for not having informed them about Plan Z earlier, but now asked for their endorsement. If they agreed, he would send the proposal to Hitler immediately, and would meet with him as soon as possible.

Chamberlain briefly discussed the alternative of a four-power meeting to resolve the issue, but he dismissed it as being unattractive to Germany. What Chamberlain left unspoken, of course, was that an international meeting would greatly diminish the drama of his visit and the acclaim that he would reap as Europe's peacemaker.

But what, Chamberlain asked, should be done about a Czech state stripped of its natural and defensible frontiers? He proposed a guarantee of the diminished country by "France, Russia, Germany, and ourselves," although he admitted that the rump state "would be a helpless little strip of territory liable at any moment to be gobbled up by Germany." Nonetheless, he hoped that the multipower guarantee would be a "deterrent" to such aggression.

Thus cabinet members heard that Chamberlain was willing to guarantee the integrity of an indefensible rather than a defensible Czechoslovakia. And he was willing to do so by entering into an agreement with countries including Russia, whose proffered assistance to an intact Czechoslovakia he had continuously rejected. How did Chamberlain's listeners make sense of this position? The prime minister tipped his hand when he said that "the inducement to be held out to Herr Hitler in the proposed negotiations was the chance of securing better relations between Germany and England." This was Chamberlain's "general settlement" in slightly different garb. Everything else—dignity, plebiscite, Czechoslovakia itself—was to be sacrificed to this hope.

The mood of the cabinet was grim, and discussion was desultory. Most members approved of Plan Z. Debate revolved around when and under what conditions the plebiscite should be held. Each minister seemed to have a different idea of what would constitute the most seemly form of dismemberment. Chamberlain promised to discuss these with Hitler, and to try to get the führer to agree to the least precipitous and disruptive plan.

Only Duff Cooper issued a fundamental challenge to Chamberlain. In the view of the First Lord of the Admiralty, "the choice was not between war and a plebiscite, but between war now and war later." In the face of the defeatism that permeated cabinet debates, Duff Cooper said that he "was confident that if we went to war we should win." However, even he cautiously endorsed Plan Z. Chamberlain thanked his colleagues for their support.

That night the prime minister sent Hitler the following telephone message: "I propose to come over at once to see you with a view to try to find a peaceful solution. I propose to come across by air and am ready to start tomorrow." At least Chamberlain had been able to spring a surprise. When he received the message, Hitler said, *"Ich bin vom Himmel gefallen"* ("I was flabbergasted").[86] The führer and the prime minister decided that the best place for a meeting would be Hitler's house, the Berghof, in Berchtesgaden.

SEPTEMBER 14, BERLIN, ARMY HEADQUARTERS

Halder learned that Hitler had returned to Berlin the previous day. Witzleben, to whom he had delegated responsibility for planning the logistics of the coup, appeared at his office and demanded, "Right. Then do I get the signal from you to strike or not?" But while Halder and Witzleben were discussing how long it would take before final preparations for a coup could be completed, they heard the news that Chamberlain had announced that he was coming to meet Hitler.[87]

SEPTEMBER 14, BERLIN, ABWEHR HEADQUARTERS

Admiral Canaris was sitting at dinner with three young colleagues when a message came from the War Ministry. Chamberlain would fly to Germany the next day to meet Hitler at Berchtesgaden. The admiral was stunned. He carefully laid down his knife and fork.

"What he—visit that man?" He was utterly distracted and unable to continue his meal. He excused himself and retired for the evening. Had he completely misjudged the British? Had it been a terrible mistake to play his hand so openly?[88]

SEPTEMBER 14, BERLIN, FOREIGN MINISTRY

When he heard about Chamberlain's impending trip, Erich Kordt remembered, "I did not expect that. The suspense had increased every hour. Hopes and disappointments quickly followed one another." What did Chamberlain think he would accomplish? He might temporarily save the peace, but didn't he understand that lasting peace could occur only through Hitler's fall? Surely the sight of the prime minister coming to Germany to beg for peace would "go to Hitler's head and stimulate him to engage in even further adventures."[89]

SEPTEMBER 15, LONDON, HESTON AIRPORT

Both Theo Kordt and Lord Halifax had come to see Chamberlain off. Although the prime minister was a sixty-nine-year-old taking his first airplane ride, he was in high spirits. Theo "wished him luck for his big task." At that point Theo was still convinced that Chamberlain was going to talk tough to Hitler. In fact Sir Horace Wilson had even asked Theo to draft a letter on behalf of the German opposition that the prime minister could send to Hitler "that would open the eyes of the German people." The letter would first be delivered to Hitler and then published. However, as Chamberlain's plane lifted off, Lord Halifax sidled up to Theo and took him by the arm. "We decided differently and thought a personal meeting with Chamberlain would be better," he whispered confidentially.[90]

SEPTEMBER 15, LONDON, MORPETH MANSIONS

As Chamberlain was departing in high spirits for his meeting with Hitler, Churchill was ruminating on the forthcoming meeting in rather a different mood. That morning the *Daily Telegraph* had carried his column on the possibility of war over Czechoslovakia. In contrast to Chamberlain's belief that the Czechs would be unable to defend themselves, Churchill thought that the Czechs would stand fast, inflicting substantial casualties on the invaders. The Germans would properly be blamed for the conflict, and the democracies would rally to the defense of Czechoslovakia. "From the moment the first shot is fired," Churchill predicted, "and the German troops attempt to cross the Czechoslovakian frontier, the whole scene will be transformed, and a roar of fury will arise from the free peoples of the world, which will proclaim nothing less than a crusade against the aggressor."[91]

While the sight of the prime minister flying to Germany to preserve the peace relieved other Britons, Churchill saw only humiliation and futility. From his modest flat in London's Morpeth Mansions, he wrote to his friend Lord Moyne, "We seem to be very near the bleak choice between War and Shame. My feeling is that we shall choose

Shame, and then have war thrown in a little later on even more adverse terms than the present."[92]

Oster and the conspirators were horrified to discover that Chamberlain was flying to meet Hitler. "We were struck dumb at first," Gisevius said. But they soon recovered, consoling themselves with the hope that the British were playing a game, wanting to show that Hitler was "glaringly in the wrong" before declaring war. Gisevius noted, "In all seriousness, we imagined that the chief danger for us lay in the possibility that not Chamberlain but Hitler might back down."[93]

When they realized later that day that Chamberlain had come not to frighten Hitler but to negotiate with him, the conspirators felt utterly defeated. As Gisevius recalled, "we bowed our heads in despair. To all appearances it was all up with our revolt."

Oster had worked quickly and efficiently in putting together the preparations for the putsch. In Halder he had an outraged and apparently engaged leader who was well placed at the center and top of the army hierarchy. Brauchitsch was a question mark, but at least he wasn't hostile. Witzleben and Brockdorff were the most vital conspirators, for it was their troops that would spearhead the seizure of power in Berlin. In the provinces the conspirators had strong supporters like Generals Adam, Hoepner, and Hase. And through the Kordt brothers and Weizsäcker, they had access to foreign information and crucial ties to the British.

The British represented a continuing source of anxiety for Oster. His various emissaries had received polite hearings, but no one except Churchill, who was out of power, had actually expressed support. While Oster and a few of his fellow conspirators would have been willing to initiate the coup regardless of the actions of the British, Halder refused to act unless the British stood up to Hitler. Oster knew that if he wanted to keep Halder committed to the conspiracy, and have a chance to bring in Brauchitsch, he would have to agree to that condition.

4

Hitler's Knife at Chamberlain's Throat: Climax

September 15, 1938, saw the beginning of a phenomenon hitherto unknown in world affairs: instant, face-to-face, crisis diplomacy, made possible by the airplane. British newspapers were virtually unanimous in praising the prime minister's daring initiative. The *Daily Herald* spoke for the British press when it called the upcoming conference at Berchtesgaden "an effort to stave off war which has seemed to be growing dreadfully near and, as such, it must win the sympathy of opinion everywhere, irrespective of Party." Theo Kordt, in an official dispatch back to Berlin, commented on the dramatically changed mood in London: "Until evening the entire British population was sunk in deep depression called forth by the decision, no less serious, to take up arms under certain circumstances. Now things have taken a completely unexpected turn which offered the hope of a peaceful settlement."[1]

The announcement of Chamberlain's impending visit had precisely the opposite effect on the conspirators. Their putsch was derailed, perhaps permanently. In tying their plans to the willingness of the

British to confront Hitler over the Sudetenland, they had made their plot hostage to the twists and turns of negotiations between Chamberlain and Hitler. As succeeding days would show, it would be a tumultuous ride.

SEPTEMBER 16, LONDON, HESTON AIRPORT

Returning from Berchtesgaden, Chamberlain deplaned in the early evening at Heston, and was immediately surrounded by reporters eager to know the results of his mission. Chamberlain said only that he had had "a great time," and that he would resume discussions with Hitler in a few days after consulting with Lord Runciman and his cabinet colleagues.[2] A waiting car carried Chamberlain to Downing Street, where he met with the "inner cabinet" of Halifax, Sir John Simon, and Home Secretary Sir Samuel Hoare at 6:00 that night. In his debriefing he told them that the only chance of avoiding war was to agree to Hitler's demand for an immediate plebiscite. While all three supported Chamberlain, they knew that their cabinet colleagues might not be as receptive to the prime minister's message the next morning.

SEPTEMBER 17, LONDON, NUMBER 10 DOWNING STREET

The cabinet gathered at 11:00 A.M. After a confusing report by Lord Runciman, Chamberlain addressed the twenty-one members about his trip to Berchtesgaden for nearly two hours. He had traveled most of the day, he said, and had arrived at the Berghof at 5 P.M. Accompanied only by Hitler's interpreter, he and the führer had retired to the small study in which Hitler had received Halifax ten months earlier.

Chamberlain thought that his personal rapport with Hitler had been excellent. Although the prime minister thought the führer "the commonest little dog," he nevertheless had seen "no signs of insanity but many of excitement." He had formed the opinion that "Herr Hitler's objectives were strictly limited," and that if he pledged peace, "Herr

Hitler would be better than his word."[3] Like a giddy schoolgirl confiding to her closest pals, he reported that he had heard that "the Führer had been most favourably impressed [by me]." Chamberlain reminded his listeners that "negotiations depended mainly upon personal contacts."

Despite this glowing appraisal of his personal rapport with Hitler, it quickly became clear that the meeting had not gone according to Chamberlain's plans. When he had suggested that they begin by discussing "a new understanding between England and Germany," Hitler had rebuffed him and said that they must first discuss the Sudeten issue. Hitler had spoken bluntly. Three million Sudeten Germans were outside the Reich, and they must come in. "If necessary," the Führer had continued, he "would run the risk of a world war to bring them in." His voice rising, Hitler had said that "300 Sudeten Germans had been killed the day before." He had insisted, "All this must be solved at once."

Further, Chamberlain related, Hitler had peppered his demands with threats. "The German military machine was a terrific instrument," Hitler had said. "Any serious incident which occurred would release the spring and the pincers would close. Once the machine was put in motion, nothing could stop it." After this saber rattling, Hitler had shaken his finger at Chamberlain and said that "if the British government would not accept the principle of self-determination there was no use in pursuing the negotiations."

Chamberlain had replied that he would have to consult with his cabinet colleagues, but as for himself, "It was immaterial . . . whether the Sudeten Germans stayed in Czechoslovakia or were included in Germany." This shameful admission—made without consulting the French, the Czechs, or even the members of his own cabinet—had surely convinced Hitler, if he had any lingering doubts, that Chamberlain would surrender everything rather than go to war.

At 1:30 P.M., the cabinet adjourned for lunch. When the members reassembled at 3:00, the discussion was lively. Duff Cooper, as usual, led the charge against the prime minister. The First Lord of the Admiralty said, "It was a primary interest of this country to prevent any single power dominating Europe." Would this proposed cession pacify Germany? He "found it difficult to believe that the self-determination of the Sudeten Germans was Hitler's last aim." And, in words both thunderous and prophetic, Duff Cooper predicted that "there was no chance of peace in Europe so long as there was a Nazi regime in Germany."

Duff Cooper's statement of principle energized his junior colleagues. "Buck" de la Warr, the Lord Privy Seal, charged, "These concessions would be unfair to the Czechs and dishonourable to ourselves," and added that for his part he "was prepared to face war in order to free the world from the continual threat of ultimatums." Lord Winterton, chancellor of the Duchy of Lancaster, said, "Sometimes war had to be faced, since otherwise the alternative was to be a vassal state." Oliver Stanley, president of the Board of Trade, prophesied, "This is not the last of Herr Hitler's *coups*. The present Nazi regime could not exist without *coups*. If the choice for the [British] Government in the next few days was between surrender and fighting, we ought to fight." Minister of Health Walter Elliot also argued that if "we were faced with the alternatives of surrender or war we must choose the latter." Although Duff Cooper's contingent was outnumbered by better than 3 to 1, they had found their voices at last, and had begun to object to a dictated settlement in language that had never been used before in the Chamberlain cabinet or scarcely in Parliament itself, except by Churchill.

The majority of the cabinet supported the prime minister. They regretted Britain's inability to defend Czechoslovakia, and some even called it a "humiliation," but they found one phrase or another to justify it. Even Chamberlain's supporters, however, emphasized the importance of getting some concessions from Germany in the next round, which was scheduled to begin on September 22, at Bad Godesberg in the Rhineland, over the scope and timing of the proposed plebiscite. They knew that they had surrendered the principle, but they wanted to salvage their consciences, and their political positions, by insisting on a German *quid pro quo*, no matter how transparent.

One very reluctant participant in the debate of September 17 was Lord Halifax. He did not join in until the very end, and he did so in words that were far from his confident rhetoric of August 30. Though he admitted that the present situation contained an element of "German blackmail, . . . this should not blind us to other considerations." War would be justified, he said, if it was fought "for the greater moralities which knew no geographical boundaries," but he didn't think this one met the standard. Unlike his colleagues in the majority, he thought that self-determination was a principle of limited significance. As he put it, "It was undesirable to burn too much incense on the altar of self-determination." Nevertheless, he noted, practical politics "made it impossible to lead this country into war against this principle." These statements persuaded nobody, but in their flaccid arguments and moral ambiguity were undercurrents of doubt about Chamberlain's program.

Hitler's truculence at Berchtesgaden had sowed seeds of doubt throughout the cabinet. For the moment Chamberlain still commanded a majority, but even his supporters wanted some indication that Hitler would be reasonable. And in the voices of Duff Cooper and his contingent and, most ominously, of Lord Halifax, can be heard the first intimations of a mutiny.

September 17, the Berghof

Hitler paced the floor as he spoke to his favorite English journalist, G. Ward Price, the Berlin correspondent for the *Daily Mail*. It was through Price that Hitler liked to send his messages to the British public. He began by declaring his respect for Neville Chamberlain, who had just visited him: "I am convinced of Mr. Chamberlain's sincerity and goodwill," the führer stated with apparent openness.

Then Hitler turned to his real purpose, spending much of the interview ranting about the Czechs: "The creation of this heterogeneous Czechoslovak Republic after the war was lunacy," he charged. "To set an intellectually inferior handful of Czechs to rule over minorities like the Germans," he observed, "with a thousand years of culture behind them, was a work of folly and ignorance."[4] Hitler signaled a warning to the Czechs in the form of a little history lesson he gave Price. It focused on the Czechs' last period of independence, in the fifteenth century: "Once indeed, during the Hussite wars the

Czechs had gained a temporary independence," the führer intoned. "They used it like Bolsheviks, burning and ravaging, until the Germans rose and crushed them."[5] Price understood the import of Hitler's words and politely excused himself to compose his story.

SEPTEMBER 20, BERLIN, HANS OSTER'S APARTMENT

The small group meeting in Oster's apartment constituted the "inner circle" of the conspiracy. It certainly included Oster, Friedrich Wilhelm Heinz, a close friend and Abwehr colleague of Oster's, Witzleben, and Franz Maria Liedig; probably Groscurth, Gisevius, and Dohnanyi; and perhaps Goerdeler. The conspirators had been meeting periodically throughout the month. Although they had been discouraged by Chamberlain's visit to Hitler at Berchtesgaden, the fact that the two leaders had not come to an agreement gave them hope that the putsch might still occur.

One of the items almost surely discussed was the political structure to follow a coup. Liedig claimed that since he joined the Abwehr in 1936, he had been busy with political plans, including "drafts of proclamations, temporary government orders, a temporary constitution, a voting system, and so on."[6] While the 1938 conspirators were never as obsessed as later conspirators with the minutiae of hypothetical postcoup governments, they did endorse a revival of the monarchy under Prince Wilhelm, the son of the crown prince, whom they respected as "a very upright, sincere and courageous soldier." To broaden the appeal of a new government, Heinz had been in touch with Wilhelm Leuschner, a trade union leader and minister, and his assistant, Hermann Maass, formerly a youth leader. While they understood the importance of a monarch as a symbol of German unity that had been lost in 1918, they also appreciated the importance of a democratic political base and a constitution. In short, the conspirators had in mind a British-style constitutional monarchy as their model government.[7]

Then they turned to an even more difficult problem. With Gisevius and Brockdorff having planned the police and military actions previously, the men in Oster's apartment focused on what to do with Hitler, a touchy issue that had dogged their discussions throughout the month.

During the second week of September, Oster had introduced Witzleben to Heinz, one of Oster's acquaintances from a darker tradition in recent German history. Heinz had been a teenage soldier during World War I. After demobilization, he had moved into the right-wing terrorist organizations that had proliferated in Germany in the 1920s. Heinz was implicated in, although never convicted of, several murders carried out by *Freikorps* organizations of which he was a member. He made a reputation as an agitator and polemicist for several political groups, including the Nazis. However, after 1933, a falling-out with his Nazi superiors had turned him into a hunted man and a dedicated anti-Nazi.[8]

Oster and Witzleben soon found use for Heinz in the developing plot. The general had ordered Heinz to assemble a raiding party (*Stosstrupp*) that would form an armed escort for Witzleben when he went to the Reich Chancellery to arrest Hitler. Heinz had done so, drawing on twenty or thirty active Wehrmacht officers, a sprinkling of student and labor leaders, and some of Heinz's old contacts in the "*Stahlhelm*," a right-wing and often violent World War I veterans' group.[9] In all, Heinz's group consisted of fifty to sixty anti-Nazis from a wide variety of opposition factions, including a few, like Heinz himself, who had past connections to right-wing terrorist organizations. As a criterion for selection, this symbolic diversity was as important to Heinz and Oster as the commandos' willingness to fight.

Oster's inner circle reluctantly agreed that Hitler's arrest would probably entail bloodshed, but they were unwilling to endorse assassination openly. There were some conspirators, like Beck, Goerdeler, and Canaris, who would have been appalled by the cold-blooded killing of Hitler. Beck had said, "Assassination is still murder. . . . Our new movement and our work needs to be focused on a better Germany that is not sullied by a murder at the beginning of its history." Goerdeler agreed. He argued that "a nation should be maintained that is built on the ethics of the Ten Commandments. This nation must not begin by flouting the

Fifth Commandment."[10] And, despite his earlier fantasies about Hitler meeting a fatal accident, Halder also favored his arrest, not his assassination.[11] With some of his main supporters expressing such strong opinions, Oster felt that he could not risk a breakdown of unity over this issue.

The group discussed Hitler's protection. In the Reich Chancellery itself, the SS troop from the Leibstandarte Adolf Hitler consisted of thirty-nine men commanded by three officers, but divided into three shifts, so that only twelve to fifteen men were on duty at any moment. The conspirators knew that there was one security guard day and night at the main entrance at Wilhelmstrasse 78, as well as at the entrance to Hitler's residence at Number 77. Another soldier patrolled the yard with a dog at night. There were single armed guards posted at the entrance to the führer's adjutant's office, the anteroom to the kitchen, the front yard, and the garage ramp on Hermann Göring Strasse. There were about four "special security service" officers who would sometimes work as reception officers. Altogether this was a significant but not impenetrable security force, which the conspirators believed could be surprised and overwhelmed by Heinz's raiders.[12]

The conspirators dispersed, having agreed that after Hitler had issued the attack order on Czechoslovakia, he would be arrested and transported to a secure location, where he would await trial. After the others had left the apartment Heinz stayed behind. He told Oster bluntly that Hitler needed to be killed: "A Hitler alive is stronger than all of our divisions." Oster had already come to the same conclusion, and the two men discussed how they would accomplish it. They decided that, regardless of whether Hitler's bodyguard offered resistance, the raiders would begin shooting, and in the confusion Hitler would be killed. Heinz understood that this plan constituted a "conspiracy within the conspiracy," and must be kept secret even from their close associates. Both Oster and Heinz were convinced that killing Hitler was the only way the coup could succeed.[13]

It is one of the ironies of history that Hitler's assassin was to be Heinz, a man who had much in common with his intended target. Both were misogynist and anti-Semitic proponents of violence. Like

Hitler, Heinz glorified war, and both had found the companionship in the trenches that had been missing from other parts of their lives. Heinz, like Hitler, had been nurtured in the nest of right-wing terrorist organizations that had befouled postwar Germany. Heinz had also been fired from his job as a journalist for the Nazi Party for being too irascible and violent. Oster had found the right man for the job.

<hr>

SEPTEMBER 21, LONDON, NUMBER 10 DOWNING STREET

<hr>

The cabinet gathered at 3 P.M. to hear Halifax's recapitulation of the events of the last two days. When Chamberlain had returned from Berchtesgaden, the British had conferred with their French allies. They had hammered out an Anglo-French plan, which basically gave Hitler everything he had demanded, and then the British had pressed the Czechs to agree to it. The Czechs had been understandably unhappy. The plan called for them to agree to surrender large parts of their country to Germany without knowing in advance exactly which parts those would be. When the Czechs balked, the British and French had let them know that if they didn't accept the plan, they would be abandoned to fight the Wehrmacht alone. These threats had worked. Bitterly resigned to accommodate their Western allies at nearly any cost, the Czechs had reluctantly agreed to accept the Anglo-French plan.

In the meantime Hitler had turned up the fires a little higher by ordering Henlein's *Freikorps* to occupy several towns in the Sudetenland. At the request of his British allies, President Beneš had not expelled them by force, so as to avoid an "incident" of the sort to which Hitler had alluded in his conversation with Chamberlain.

Halifax announced that while the ministers of His Majesty's Government had been scurrying about, gaining the proper assent to the demands that Hitler had made at Berchtesgaden, the Germans were becoming "restive" about the time for the follow-up meeting. Hitler wanted it to start the next day, September 22, at the Rhineland town of Bad Godesberg, and Halifax and Chamberlain had agreed to these arrangements.

Finally Halifax told the cabinet that the Hungarian and Polish governments were also pressing claims against Czechoslovakia on behalf of *their* minorities. Halifax had been trying to put them off, pleading that the British had first to deal with the more pressing issue of the Sudeten Germans. Chamberlain added that while Hitler had initially disavowed any interest in the fates of other minorities in Czechoslovakia, "representatives of Hungary and Poland had both visited Berchtesgaden in the last few days, and we had no very precise account of what had passed."[14] The prime minister said that he had to discuss with the cabinet his response if Hitler should insist on "self-determination" for other minorities. Chamberlain felt that if the führer did so, he should break off the talks immediately and "return home to consult his colleagues." Sir Samuel Hoare agreed with the prime minister, saying "this issue would be a test of Herr Hitler's sincerity." Halifax concurred, "the Prime Minister should refuse to yield to pressure from Herr Hitler" if he raised the question of the other minorities.

Everyone in the room understood the significance of the issue. Malcolm Macdonald, colonial secretary, said it explicitly: "If Herr Hitler pressed for an immediate settlement of these claims, it would show that his interests lay, not in the German race, but in an attempt to dominate Europe." The question of Hitler's ultimate aims hung over the room like a pall. If he wanted only reunification of ethnic Germans living outside the Reich, they would swallow hard and accede to that as a cost of avoiding war. But if Hitler's *real* goal was to dismember Czechoslovakia as the first step to dominating Europe, they would have to resist.

In trying to anticipate the course of the upcoming negotiations, the cabinet considered two other issues. Would Germany agree to a "guarantee" by Britain, France, and the Soviet Union of Czechoslovakia shorn of the Sudetenland? And what if Hitler should insist on immediate occupation of the Sudetenland by German troops, pending a formal transfer of territory and population? Duff Cooper spoke most vociferously against letting German troops enter Czechoslovakia. "It would end by their overrunning the whole country, . . . and they would not stop until they were in Prague." Even Sir Kingsley Wood, secretary for air and one of Chamberlain's staunchest supporters, allowed that "it would be a shock to public opinion if German troops were allowed to enter Czechoslovakia immediately."

All three of these issues—Hungarian and Polish minorities, a multinational guarantee to Czechoslovakia, and immediate occupation by German troops—were made into tests of Hitler's intentions, and his willingness to negotiate rather than demand changes at the point of a gun. The cabinet decided that, should Hitler balk at the British position on *any one* of these issues, Chamberlain would immediately break off talks and return to London.[15]

SEPTEMBER 22, LONDON, HESTON AIRPORT

Chamberlain was seen off by Halifax, Theo Kordt, and the dignitaries and diplomats who had wished him well just a week earlier. As the journalists waited with poised pencils, Chamberlain mouthed his usual bland phrases: "A peaceful solution of the Czechoslovakia problem is an essential preliminary to a better understanding between the British and German peoples," and "European peace is what I am aiming at." But the public mood had changed. There was skepticism in the air about whether Hitler was negotiating seriously. Geoffrey Dawson of *The Times* predicted to Theo Kordt that "if Chamberlain were to return without an understanding based on the Anglo-French plan, public opinion would turn against him." Kordt, who had noted the abrupt change in mood before Chamberlain had departed for Berchtesgaden seven days earlier, concurred. He wrote to Berlin that "opposition to Chamberlain's policy is increasing."[16]

SEPTEMBER 22, BAD GODESBERG

Chamberlain and his entourage arrived at the resort town of Bad Godesberg to begin the next round of negotiations, which they believed would resolve the details of a plebiscite in the Sudeten German regions.

They were staying at the Kurhotel in Petersberg, a village directly across the Rhine, and had to be ferried back and forth across the river for their meetings. The German delegation was staying in Hitler's favorite hotel, the Rheinhotel Dreesen in Bad Godesberg, and included Joachim von Ribbentrop, Paul Schmidt, his translator, and Eduard Brücklmeier, a diplomat who also served as a translator. Unknown to Hitler and Ribbentrop, Schmidt and Brücklmeier were informants who kept Erich Kordt and Oster apprised of the proceedings.[17]

SEPTEMBER 23, BERLIN, FOREIGN MINISTRY

In the afternoon Brücklmeier telephoned Erich Kordt, who had Oster in his office, to tell him the wording of the letters that had been exchanged earlier that day between Chamberlain and Hitler. When he heard the news, Oster was energized. "Now we have, thank God, clear evidence that Hitler wants war under almost any circumstances. There is no turning back." But he added to Kordt, "Do everything you can to get Hitler to return to Berlin. The bird must return to the cage."[18] Oster's anxiety was understandable: He knew that the putsch depended on the führer's presence in Berlin, where he had spent very few days since the end of August.

Oster was elated when Kordt told him that Hitler would travel to Berlin the next day. Oster agreed that Fritz von der Schulenburg would keep Kordt informed of the coup plans. As he left the room Oster asked, "Can you get us an exact plan of the Reich Chancellery?" Kordt replied that he already had one. "For the first time in weeks," Kordt said, "I had a feeling of relief." Oster left to pass on the news from Bad Godesberg to the other conspirators.

SEPTEMBER 24, BERLIN

While the conspirators expected the worst from Chamberlain's second trip to Germany, "we could scarcely believe our ears at the strange

Hans Oster (1888–1945).

Oster and his horse, Bobby.

Oster.

Lord Halifax in his office.

Hans von Dohnanyi (1902–1945).

Ludwig Beck (1880–1944).

Werner von Fritsch (1881–1939).

Erwin von Witzleben (1881–1944).

Generals Ludwig Beck (*left*) and Werner
von Fritsch on army maneuvers in
northeastern Germany, 1937.

Bernhard Lichtenberg
(1875–1943).

Friedrich Wilhelm Heinz
(1899–1968).

Army Command Headquarters in
Bendlerstrasse.

Werner von Fritsch (*left*), Hans Oster (*center*),
and Friedrich Wilhelm Heinz in the Alps.

Carl Friedrich Goerdeler (1884–1945).

Erich Kordt (1903–1970).

Theo Kordt (*left*) accompanies Neville Chamberlain as the latter is about to board an airplane on one of his three trips to Germany, September 1938.

Winston Churchill as prime minister, visiting an armaments factory in the north of England, summer 1940.

reports that came from Godesberg," according to Gisevius.[19] Hitler and Chamberlain sat in their respective hotels on either side of the Rhine like two lovers sulking after a spat. Soon the rumors were confirmed: "Hitler had retracted the proposals he himself had made at Berchtesgaden and was now making new demands, such monstrous demands that they would certainly be too much even for Chamberlain." Gisevius continued, as news leaked out, "Disappointment, indignation, and panic spread throughout Germany. Never before had the Germans spoken so freely and vituperatively. Strangers talked to one another on the streets. The fearful shock could be read plainly in people's faces. This time Hitler had gone too far."[20]

Encouraged by this unexpected turn of events, the conspirators began to reactivate their plans. "Even Brauchitsch mumbled grim threats," Gisevius noted.[21] The commander in chief had been most reluctant to discuss a coup, but the turn of events at Bad Godesberg seemed to awaken even this timid lion. Erich Kordt remembered that "at the Godesberg Conference Colonel Oster informed me that von Brauchitsch was ready to participate in a revolt."[22] That was an essential commitment. Oster had been trying to woo Brauchitsch to the conspiracy since late August, when he had sent Kordt to approach him. Now, at last, the commander in chief of the army had declared that he would join the plot.

Gisevius nearly quarreled with Oster "for the first and last time in this life" because of Oster's momentary pessimism. Despite his elation of the previous day, Oster believed that the British would ultimately yield to Hitler. "I told him that he deserved a post in the propaganda ministry."[23]

SEPTEMBER 24, LONDON, NUMBER 10 DOWNING STREET, 5:30 P.M.

Chamberlain, who had returned from Bad Godesberg earlier in the day, called together his cabinet to brief them about his meeting with Hitler. Word had leaked out to some cabinet members and to the press that the meeting in Germany had not gone at all well. While Chamberlain had been negotiating in Bad Godesberg, Halifax had sent him a message that "the great mass of public opinion [seemed] to be hardening" toward further concessions to Hitler. Indeed, Halifax had told Chamberlain bluntly that if Hitler, despite the enormous concessions in the Anglo-French plan, decided to go to war over Czechoslovakia, that he, the cabinet, and the British people would regard it as "an unpardonable crime against humanity."[24]

The mood was tense as Chamberlain recounted his meeting with Hitler in the führer's hotel with their respective interpreters. Hitler had listened quietly while Chamberlain had explained his success in persuading his cabinet, the French, and even the Czechs to accept the demands that Hitler had made at Berchtesgaden. When Hitler finally spoke, he had said that "he was sorry, since the proposals were not acceptable to him."[25] Chamberlain noted that Hitler's blunt rejection of his *own* demands made at Berchtesgaden "had been a considerable shock."

Hitler had presented his new agenda. German troops must occupy all of the Sudetenland immediately, beginning no later than Monday, September 26, to be completed two days later. The date for a plebiscite—the first time that Hitler had actually used that term—could be set for sometime in November. Furthermore the claims of the Hungarian and Polish minorities must be addressed before Hitler would consider signing a treaty of nonaggression with a restructured Czechoslovakia.

Chamberlain had said that this was "an impossible proposal." Public opinion in Britain would regard the immediate German occupation of the Sudetenland as equivalent to "the seizure of a conquered territory." Hitler had said that he didn't see why, and suggested that they adjourn to a room downstairs to consult a map that delineated the areas to be occupied.

At this point Hitler had raised two of the issues—immediate German occupation and Hungarian and Polish minority claims—that the cabinet had designated as deal-breakers. Chamberlain had pledged—indeed, he had originally *suggested*—that if Hitler insisted on either of these, he would return to London to consult with his colleagues rather than continue the conversation. But continue he now did. Joined by Sir Nevile Henderson, Sir Horace Wilson, and Ribbentrop, Hitler and Chamberlain had hunched over maps spread out on a long table. Rather than refuse to discuss the issue further, Chamberlain said that he had been "relieved to notice that the area on this map did not appear very different from the 50% area marked on the maps which had been examined in this country."

Hitler had orchestrated the scene beautifully. Messengers had periodically interrupted the discussions, bringing word of still another Czech outrage perpetrated against innocent Sudeten Germans. Hitler had cast himself as an avenging angel, struggling to control his righteous fury. When the prime minister had asked Hitler to control the Sudeten Germans, the führer had shouted that they "were unorganized, their leaders had been arrested or had fled the country, and he could not control them." (Chamberlain described Hitler's conversational style as "discursive.") On this threatening note, Hitler had adjourned the meeting until the next day.

Chamberlain had requested that Hitler put his new demands in writing, which he did. The two delegations, having exchanged proposals, met again at 10:30 P.M. on September 23. When Chamberlain reproached Hitler for having made no concessions to match his own, Hitler had claimed that *his* concession was *not* invading and overrunning Czechoslovakia! After Chamberlain objected that the German demands constituted an ultimatum rather than a memorandum, Hitler had genially agreed to two changes. The words "German demands" would be replaced by "German proposals" in an otherwise unchanged document, and the completion date for Czech evacuation and German occupation of the Sudetenland would be pushed back from September 28 to October 1. Hitler told Chamberlain that he was "the only man" to whom he had ever made a concession.

In evaluating the events for his cabinet, Chamberlain confessed that he had first been "indignant" when he had left Bad Godesberg. But now that he had had a chance to reflect on the meeting, he "did not believe that Herr Hitler thought that he was departing in any way from the spirit of what he had agreed to at Berchtesgaden." Furthermore he was "sure that Herr Hitler was extremely anxious to secure the friendship of Great Britain." And finally Hitler had assured him that his goal was "racial unity" and "not the domination of Europe." On this last crucial question—how to evaluate Hitler's intentions—Chamberlain "believed that Herr Hitler was speaking the truth."

Chamberlain had not just taken the bait, he had helped Hitler put in the hook. "Herr Hitler had also said that, once the present question had been settled, he had no more territorial ambitions in Europe. He had also said that if the present question could be settled peaceably, it might be a turning-point in Anglo-German relations." Chamberlain's enormous capacity for self-delusion was reflected perfectly in his evaluation of his relationship with Hitler: "A peaceful settlement of Europe depended upon an Anglo-German understanding." Chamberlain was certain that "he had now established an influence over Herr Hitler, and that the latter trusted him and was willing to work with him."

Chamberlain asked the cabinet members to think about whether the differences between the proposals Hitler had made at Berchtesgaden and Bad Godesberg "justified us in going to war." For himself, he thought not. "That morning," Chamberlain had "flown up the Thames over London, . . . and had imagined a German bomber flying the same course."[26] Chamberlain thought of the terrible destruction that would result and "felt that we were in no position to justify waging a war to-day to prevent a war hereafter." He said that in his opinion, there was no choice but to allow the Germans to occupy the Sudetenland since Britain did not have the force to stop them. "If we now possessed a superior force to Germany, we should probably be considering these proposals in a very different spirit." Chamberlain's tragic miscalculation of Hitler's intentions, flawed military intelligence, and deep abhorrence of war brought him to a position that can only be called abject defeatism.

Duff Cooper was the first to attack Chamberlain's assumptions. The First Lord of the Admiralty predicted "that in the event of a German attack on Czechoslovakia, he felt sure that public opinion would bring about a position in which we should have to intervene in the war." However, he "feared that unless we acted promptly that intervention might come too late to be effective." In contrast to the prime minister, Duff Cooper believed that "we could place no confidence whatsoever in Hitler's promises." Furthermore he "was certain that Herr Hitler would not stop at any frontier which might result from the proposed settlement." If war were to break out, Britain was not powerless. It could use the strategy of naval blockade, which had proved to be so effective toward the end of the previous war.

The remainder of the discussion revolved around whether war was sufficiently close that His Majesty's Government should order mobilization of the armed forces. Duff Cooper, joined by William S. Morrison, minister of Agriculture and Fisheries, Winterton, and Hore-Belisha, argued that it was, while Chamberlain, Halifax, and Sir John Simon preferred to await developments. Engaged at sword's point over the most crucial issue of national security—war preparations—the cabinet members decided to sleep on it and adjourn until the next morning.

SEPTEMBER 25, LONDON, NUMBER 10 DOWNING STREET, 10:30 A.M.

In the busy days of September 1938, Lord Halifax could often be seen walking—if the weather was fine—from his home to his office with Sir Alexander Cadogan, permanent undersecretary of the Foreign Office. They made an odd pair, Halifax tall and lanky, striding purposefully, while the much more compact "Alec" scurried to keep up the pace. Mutt-and-Jeff appearances notwithstanding, the two Old Etonians liked and trusted each other and candidly shared opinions about professional matters.

On the night of September 24–25 Halifax had suffered a rare bout of insomnia because of a recent conversation with Cadogan. As he told his friend the next day, "Alec, I'm very angry with you. You gave me a sleepless night. I woke at 1 and never got to sleep again."[27] Cadogan absorbed this mock dressing-down with the well-bred manners of a lifelong civil servant, but secretly he was pleased that he had caused Halifax disquiet. He had been deeply troubled by Chamberlain's description of the Bad Godesberg conference. Cadogan confided his shock to his diary:

A week ago when we moved (or were pushed) from "autonomy" to cession, many of us found great difficulty in the idea of ceding people to Nazi Germany. We salved our consciences (at least I did) by stipulating it must be an "orderly" cession—i.e., under international supervision, with safeguards for exchange of population, compensations &c. Now Hitler says that he must march into the whole area at once (to keep order!) and the safeguards—and plebiscites! can be held after![28]

Cadogan's incredulity and anger had grown as the day wore on. "I was completely horrified—he [Chamberlain] was quite calmly for total surrender. More horrified still to find that Hitler has evidently hypnotised him to a point. Still more horrified to find P[rime] M[inister] has hypnotised H[alifax] who capitulates totally." He went so far as to characterize his boss's position as "*défaitiste*-pacifist." Cadogan clearly understood that Britain and France were not completely ready to make war in September 1938, but he also understood the consequences of giving in to Hitler's bullying: "I know we and they [the French] are in no condition to fight: but I'd rather be beat than dishonoured." After the cabinet meeting ended that night, Cadogan "drove [Halifax] home and gave him a bit of my mind." "But," he added, it "didn't seem to shake him."[29]

However, although Cadogan didn't know it at the time, his pointed words had struck their mark. Halifax had had an epiphany. In the throes of his late-night insomnia, Halifax said later to Cadogan, "I came to the conclusion that you were right." Since his meeting with Hitler in November 1937, Halifax had been a loyal advocate of Chamberlain's policy of appeasement. He had listened to Churchill's denunciations of Hitler's aggression and his calls for a "Grand Alliance" against Nazi Germany, and he had received the urgent, whispered warnings of the German conspirators Ewald von Kleist-Schmenzin and Theo Kordt that Hitler was bent on war. Nevertheless his faith, while wavering, was unbroken.

As of 10:30 P.M. on September 24, Halifax still supported Chamberlain's grand vision: If Britain and France could find the right package of concessions, Hitler would be satisfied, and Germany—even a

Germany ruled by Nazis—would rejoin the community of peace-seeking European nations. Prodded by Cadogan's doubts, however, Halifax saw the world differently the next morning. Hitler had shown that he could not be trusted. His Majesty's Government clearly was negotiating with a man who was certainly untruthful and possibly insane. If Hitler couldn't be counted on to keep his word, Halifax concluded, the policy of appeasement was fatally flawed.

The minutes of the cabinet meeting of the morning of September 25 reveal a changed foreign secretary. Halifax admitted to his colleagues that he had been wavering on the Sudeten question for the last week, and that he had especially "found his opinion changing in the last day or so." He spoke carefully about Hitler's demands and assurances. "Much, of course, turned on Herr Hitler's future intentions and on the Anglo-German *rapprochement* of which the Prime Minister had spoken on Saturday." Nevertheless, "[Halifax] could not rid his mind of the fact that Herr Hitler had given us nothing and that he was dictating terms, just as though he had won a war but without having had to fight."[30]

The connection between the two parts of Halifax's statement was crucial. He was asking his colleagues, "Can you trust a man who negotiates like he is dictating a Carthaginian peace to keep the promises that he has made about the future?" Halifax answered for himself with startling bluntness: "The ultimate end which he (Lord Halifax) wished to see accomplished [was] the destruction of Nazism. So long as Nazism lasted, peace would be uncertain." In its unqualified statement of principle—so unlike Halifax!—one can scarcely imagine a more far-reaching disavowal of appeasement. Furthermore, in a veiled but unmistakable reference to the German conspirators and their plans, Halifax speculated whether if Hitler "was driven to war the result might be to help bring down the regime."[31] War—or at least the declaration of war—might achieve the ends that appeasement could not.

One of the less attractive tasks that had fallen to the foreign secretary over the past several months had been forcing the Czechs to accept the developing dismemberment of their country. Every time that Hitler had made fresh demands, Halifax had had to pressure the Czechs to accept them. Now, Halifax said, he would no longer do so. In speaking of Hitler's latest ultimatum, "We should lay the case before them [the Czechs]. If they rejected it," Halifax predicted, "France would join in." Halifax added ominously, "if France went in we should join with them." This amounted to the very sort of pledge that he and Chamberlain had been running from for the past year. If the Czechs resisted a German attack, and the French honored their treaty commitments to them, Britain should also join the conflict. Halifax was saying that if it took a world war to stop Nazi aggression, then so be it. His statement traced the path that had been laid out in earlier cabinet discussions by Duff Cooper, but in its uncompromising tone was positively Churchillian.

Although he admitted that he "had worked most closely with the prime minister throughout the long crisis," Halifax was "not quite sure that their minds were still altogether at one."[32] Camouflaged by this massive understatement was the fact that Halifax was raising the banner of revolt against the prime minister just days before Hitler's threatened attack on Czechoslovakia. Chamberlain recognized Halifax's attack for what it was. He immediately scribbled a note to him: "Your complete change of view since I saw you last night is a horrible blow to me, but of course you must form your opinions for yourself." In the same note Chamberlain referred to pending discussions with a delegation from the French cabinet that was just then traveling to London. "It remains however to see what the French say. If they say they will go in [i.e., defend the Czechs], thereby dragging us in I do not think I could accept responsibility for the decision." This last phrase was a threat that, rather than support the French in a war over Czechoslovakia, he would resign. If Halifax was going to hurl thunderbolts across the cabinet table, Chamberlain would hurl them right back.

Halifax immediately replied: "I feel a brute—but I lay awake most of the night, tormenting myself and did not feel I could reach any other conclusion at this moment, on the point of coercing Cz[echoslovakia]." Chamberlain read Halifax's note and scrawled acerbically on the bottom, "Night conclusions are seldom taken in the right perspective. N.C."[33]

While the two erstwhile allies were passing notes back and forth, the debate continued. Halifax's change in position energized the cabinet dissenters—Oliver Stanley, Alfred Duff Cooper, and Leslie Hore-Belisha—who stressed the immorality of deserting Czechoslovakia; but they remained in a minority.[34] The majority, who agreed that Hitler's demands had been outrageous, nevertheless continued to support Chamberlain's policies, stressing the impracticality of defending the Czechs and the horror of getting into another world war. The cabinet, which met in both morning and afternoon sessions on September 25, failed to resolve the vast gap that Halifax had opened up between the prime minister's position and his own. Cabinet ministers, sensing that a debate on this divergence might bring the government crashing down at this critical moment, deferred making a decision about anything until they had sounded out the French.

SEPTEMBER 25, LONDON, NUMBER 10 DOWNING STREET, LATE AFTERNOON

Jan Masaryk, the Czech Ambassador to London, was the sort of charming and highly cultured European whom Englishmen found irresistible. He represented his country with intelligence, integrity, and a keen sense of humor. (He once said that his main task in London was convincing Englishmen that Czechoslovakia was a small country rather than an exotic disease.) His father, Thomas Masaryk, had been one of the political founders of the country, and Jan was proud to carry on his work.

During the morning of September 25, Masaryk had had a long telephone conversation with President Beneš in Prague. They discussed the memorandum that Hitler had presented to Chamberlain at Bad Godesberg. Although they had swallowed hard and had accepted the "Anglo-French plan" that had come out of the Berchtesgaden conference, Hitler's latest demand for immediate German occupation of strategically vital sections of their country was too much. As Beneš said, "[I]t means we put our whole state into Hitler's hands."

After the cabinet meeting adjourned late that afternoon, Masaryk found Chamberlain and Halifax together. He delivered a copy of a letter that he had written, based on his earlier conversations with his president. Of Hitler's Bad Godesberg memorandum the letter stated simply, "it is a de facto ultimatum of the sort usually presented to a vanquished nation and not a proposition to a sovereign state which has shown the greatest possible readiness to make sacrifices for the appeasement of Europe."[35] Masaryk noted that "the proposals . . . deprive us of every safeguard for our national existence." For these and many other reasons, "My government wishes to declare that Herr Hitler's demands in their present form are absolutely and unconditionally unacceptable to my Government." And in a conclusion that evoked the glorious history of his country, Masaryk wrote, "The nation of St. Wenceslas, John Hus and Thomas Masaryk will not be a nation of slaves."

The Czechs, after months of reluctant compromise, had finally taken a stand and declared that they were ready to fight. In a postscript to his letter, Masaryk reminded Chamberlain and Halifax of their moral responsibility: "We rely upon the two great Western democracies, whose wishes we have followed much against our better judgment, to stand by us in our hour of trial."

SEPTEMBER 25, LONDON, NUMBER 10 DOWNING STREET, 11:30 P.M.

The cabinet met for the third time that day to hear Chamberlain's report on his meeting with the French earlier that evening. Premier Edouard Daladier had told Chamberlain that the French cabinet had rejected the German memorandum, having come to the view that "Herr Hitler's object was to destroy Czechoslovakia and to dominate Europe."[36] The prime minister had interrogated the French at great length about their intentions if the Germans should attack Czechoslovakia. To each of these hostile cross-examinations, Daladier had answered cryptically but clearly, "France would fulfill [its] obligations." When the French premier had turned the tables on Chamberlain, and asked him about Britain's intentions, Chamberlain had refused to respond.

Duff Cooper listened, fuming, to Chamberlain's story, then went on the attack: "The British Ministers had appeared to have contested the French point of view on every point and to have allowed it to appear that they disagreed with the French Government's suggestions without making any positive contribution themselves."

Chamberlain was struggling. The Czechs and the French had both rejected Hitler's demands, Halifax had deserted him, and attacks by dissident cabinet members were becoming more pointed and personal. Still, Chamberlain was not ready to concede that war was inevitable. He asked for the cabinet's approval to send Sir Horace Wilson to Berlin carrying a personal letter from him to Hitler making a "final appeal" for peace. In order to get cabinet approval of his plan, however, Chamberlain knew that he would have to put some teeth in it. He promised that he would instruct Wilson to warn Hitler that "if this appeal was refused, France would go to war, and that if that happened it seemed certain that we should be drawn in." Duff Cooper argued for an even stronger message, but Chamberlain rebuffed him, saying that he didn't want to present Hitler with anything that "looked like a threat." The cabinet agreed. Chamberlain would have his last chance.

SEPTEMBER 26, BERLIN, ABWEHR HEADQUARTERS

Canaris chose this moment to announce the promotion of Hans Oster to second in command of the Abwehr. Henceforth he would effectively be Canaris's chief operating officer. Canaris's biographer stated the message carried by this action: "Nobody who hoped to prosper in the Abwehr could afford to bypass Oster from now on."[37] Why did Canaris do it? His relationship with Oster had been dogged by a number of differences almost from the beginning. Oster was candid to the point of indiscretion; Canaris was secretive. Oster had been a foe of the Nazis since 1934; Canaris had waffled about his support for Hitler until very recently. Oster had been using the Abwehr to plan a revolt; Canaris had offered only occasional encouragement. Now these differences were buried in their shared goal. Although Canaris preferred to remain in the background, Oster's promotion was a clear sign to other Abwehr officers that Canaris supported his plans for a coup d'état.[38]

Halifax, now at odds with Chamberlain, agreed to talk to Churchill when the latter came calling on him at the Foreign Office. Churchill at once pressed for the same statement that he had suggested on August 31, "a declaration showing the unity of sentiment and purpose between Britain, France, *and Russia*, against Hitlerite aggression."[39] With war imminent, Halifax yielded to Churchill's pleas. As Churchill remembered it, "We discussed at length and in detail a communiqué, and we seemed to be in complete agreement. Lord Halifax and I were at one." He added, somewhat disingenuously, "I certainly thought the Prime Minister was in full accord." Churchill worked on the draft throughout the afternoon with Reginald Leeper, from the News Office of the FO. "When we separated, I was satisfied and relieved," Churchill remembered.[40] The "Leeper telegram," as it is known, bears the hallmarks of Churchill's blunt, confrontational prose:

> *The German claim to the Sudeten areas has clearly been conceded by the French, British, and Czechoslovak Governments, but if in spite of all efforts made by the British Prime Minister a German attack is made upon Czechoslovakia the immediate result must be that France will be bound to come to her assistance, and Great Britain and Russia will certainly stand by France.*[41]

At 8:00 P.M., the message was approved by Halifax and released to the press. The prime minister saw it as yet another hostile volley in the growing feud with his foreign secretary. Halifax noted that "greatly to my surprise, Neville was much put out when the Communiqué appeared, and he reproached me with not having submitted it to him before publication." Halifax claimed to be perplexed about the source of Chamberlain's anger, "unless it was that he thought it 'provocative' and not fully consistent with his desire to make a further conciliatory appeal to Hitler."[42]

Sir Horace Wilson, accompanied by Nevile Henderson and Sir Ivone Kirkpatrick, first secretary at the Berlin embassy, arrived carrying his letter from Chamberlain to Hitler. Oster's plan now hung on the führer's response to this last-ditch effort to find a peaceful solution. If he rejected it, it would indicate that an invasion of Czechoslovakia would likely follow, and the conspirators would spring into action.

The tone of the meeting was not conciliatory. "Hitler was in one of his worst moods," according to Kirkpatrick. "He was only induced with much difficulty to listen to the Prime Minister's letter." Paul Schmidt remembered that "the letter produced one of the most stormy meetings that I have ever experienced."[43]

Chamberlain's letter announced that the Czechs had refused to accept Hitler's Bad Godesberg demands. Before Sir Horace could finish reading it aloud, Hitler leaped to his feet, shouting, "There's no point at all in going on with negotiations," and he rushed toward the door as if to leave. But even the führer seemed to realize the absurdity of ending the meeting by stalking out of his own office, and he returned to his seat "like a defiant boy." Schmidt remarked that Hitler "let himself go more violently than I ever saw him do during a diplomatic interview." When Wilson attempted to calm Hitler, "it only increased his fury."[44] And when Sir Horace spoke about the British desire for a peaceful solution, "Hitler pushed back his chair and smote his thigh in a gesture of suppressed rage." Kirkpatrick noted laconically that Hitler seemed "bent on having his little war."[45]

Hitler stood behind the podium at the Sportpalast on Potsdamerstrasse. The centrally located, spacious indoor hall had long been a popular venue for Hitler to address the German people. On this night he looked out at a sea of his followers, many wearing the brown or black uniforms of the SA or the SS. To his right was an enormous banner proclaiming *"Ein Volk Ein Reich Ein Führer!"* (One People One Country One Leader!). Beyond the visible audience, Hitler was speaking by radio to the people of Germany and the people of Europe. They anxiously awaited his words. Would it be peace or war?

In the audience was William L. Shirer, a Berlin-based correspondent for CBS. Shirer noted that the recent Czech and the French rejections of his demands had put Hitler into "one of the worst rages of his tumultuous life."[46] The young American journalist described Hitler's speech that night:

> *Shouting and shrieking in the worst paroxysm I had ever seen him in, he hurled a torrent of insults against "Herr Beneš," denouncing him as the "father of lies." Trying to work up his audience of fifteen thousand Bonzen [big shots] into a suitable state for war, he grimly depicted the Sudetenland as "under a Czech reign of terror," where "whole stretches of the country were depopulated, villages burned down and attempts made to smoke out Germans with hand grenades and gas."*

Shirer's journalistic skills were stretched to their limits. He was broadcasting live, trying to translate Hitler's words into English as they were being delivered above the din of an enthusiastic crowd. The führer reiterated his shopworn pledge that he had no further territorial claims in Europe after the Sudetenland. "'We want no Czechs,' he muttered contemptuously. But he must have the Sudetenland by October 1—five days hence—he said. If 'Herr Beneš' didn't turn it over to him by then, he would take it. 'It is now up to Herr Beneš! He has the choice: peace or war!'" Shirer thought, "for the first time in all the years I've observed him, Hitler seemed tonight to have completely lost control of himself."

With this direct and public challenge, Hitler had committed himself to occupy the Sudetenland by October 1. Now that the world had heard him, he could not turn back. There would not be another May 21!

Despite this belligerent performance by Hitler, the mood was festive. According to Shirer, "The men and women in the vast audience were almost good-natured, as if they didn't realize what the Leader's words meant." Also listening carefully to the führer's words, but with a similarly more ominous sense of their meaning than the Nazi functionaries in the hall, was Ian Colvin. "I noticed that his voice was uncertain, with long blank periods, caused by the mental strain of the crisis." Colvin speculated that Hitler had been shaken by the unwavering tone of the press release that had so recently been issued over the signature of Lord Halifax.[47]

SEPTEMBER 27, BERLIN, REICH CHANCELLERY

Sir Horace was invited back for one last attempt to reach an agreement with Hitler. Wilson asked Hitler to pledge to abstain from using force if the British would guarantee that the Czech evacuation would proceed quickly. Hitler angrily refused to discuss it. The Czechs' only option, he said, was to accept Germany's demands. "If the Czechs have not accepted my demands by 2 P.M. on Wednesday September 28th," Hitler said ominously, "I shall march into the Sudeten territory on October 1st with the German army."[48]

After listening quietly to Hitler's rantings, Wilson suddenly rose to his feet. In a slow cadence, "weighing each word," he delivered Chamberlain's message: "If France, in fulfillment of her treaty obligations, should become actively involved in hostilities against Germany the United Kingdom would deem itself obliged to support France."

Hitler was beside himself with fury. "It means that if France chooses to attack Germany, England feels it her duty to attack Germany also." With his voice rising, he went on, "If France and England want to unleash war, they can do so. It's a matter of complete indifference to me. I am prepared for all eventualities. I can only take note of the position. So—next week we'll all find ourselves at war with one another." That was Hitler's last word. Kirkpatrick felt that the die was cast. "As I grasped his podgy hand I felt an overwhelming sense of relief that war would come and that I should never have to see him again." Wilson left for London immediately.[49]

Erich Kordt was now in a fevered state, thinking that war would soon be declared and the coup would be launched. But his superior, State Secretary Weizsäcker, insisted that he concentrate on the Reich's diplomatic business at hand. "Chamberlain has still not gotten an answer to his letter he sent through Sir Horace," Weizsäcker said. "It is our duty as the Foreign Ministry to draft a conciliatory response. Let's try to tie Hitler down to his Sportpalast speech [in which he had said 'we want no Czechs']."

Kordt could barely keep his mind on the task. "I made objections but the State Secretary interrupted me harshly."

"And if the coup does not work out?" Weizsäcker asked. "We must have done everything on our level in order to prevent war and must not play va banque [that is, to go for all or nothing]."[50]

Earlier in the day Hitler had ordered his assault troops—some seven divisions—to be moved secretly to their attack positions along the Czech border. The political haggling during September had put the Czechs on military alert, but the führer was still hoping that the invasion, on September 30, might take the defenders by surprise. In the meantime he decided to parade the Wehrmacht through central Berlin to whip up a martial spirit among the citizenry.

Lieutenant Colonel Edgar Röhricht was about to go to the forward areas where the troops of IV Corps would launch their attack against Czechoslovakia. At the last minute General Olbricht, chief of staff to General List, the local Wehrkreis commander, told him to hold off, "in case the plans work out."[51] In fact, Olbricht said, some of the motorized units might be sent "in an entirely different direction." Puzzled by his general's orders, but remembering the visit that had been paid them by Goerdeler in the wake of the Fritsch crisis in February, Röhricht blurted out, "A coup d'état then!"

"Call it what you will," Olbricht answered. "More of an act of self-defense."

The junior officer tried to imagine what might happen if Göring's Luftwaffe began the attack on Czechoslovakia while their troops, and others, either stood fast or perhaps even took off toward Berlin. "It will be unimaginable confusion!"

Olbricht replied that that was precisely why he wanted Röhricht to help him sort everything out.

Röhricht was excited but alarmed by the prospect. "Do you think, General, that we have units in our area that are immediately willing to be thrown into war against Hitler, particularly as no one has been prepared for such a turn-out?"

Olbricht simply shrugged and said, "Who can say?"

Throughout the provinces there were undoubtedly other generals who stood fast, awaiting orders from Army Headquarters to turn their troops away from the Czech border and participate in the coup d'état. This was the fruit of plantings the previous spring, when Oster had sent out his "missionaries" to provincial troop commanders. The seedlings of discontent had been nurtured in the summer, when Beck's reasoned dissent from Hitler's aggressive policies had been disseminated at the General Staff war game and the meeting of the *Generalität* on August 4. At the time Oster thought that he had failed, but now it looked as though he had laid the groundwork for success.

SEPTEMBER 27, BERLIN, WILHELMSTRASSE

Susanne Simonis and her friends were watching the developments in Berlin with fear and trepidation. They were alarmed at Hitler's bellicose speeches and the inexorable movement toward war. Ruth Andreas-Friedrich, a young writer, and Simonis met for coffee at a café just across from the Reich Chancellery. Like their countrymen they remembered World War I. A generation earlier, there was scarcely a family in all of Germany that had not lost a son, a brother, or a father on the battlefield. During 1917–18 the British blockade had nearly starved the civilian population to death. After surrender came the chaos of a shaky new government, with rebellions from both the right and the left, quickly followed by the ruinous inflation of 1923. This was what war had brought to Germany.

The two women watched a unit of troops marching in front of the Reich Chancellery and observed the reaction of the other spectators. Andreas-Friedrich recalled the scene on that fateful day:

> *About two hundred people are standing in the square outside the Reich Chancellery . . . clustered tightly together, they are staring tensely at an uninterrupted procession from Unter den Linden past the historic balcony: cannon, baggage trucks, horses, tanks, soldiers. Soldiers without end. Steel helmets low on their foreheads, eyes rigidly to the front, they sit in their saddles, crouch on wagon seats and limbers, march heavy-footed along the asphalt.*
>
> *The balcony door upstairs opens. Hitler comes out bareheaded, his hands in the pockets of his tunic. He moves quickly to the railing. . . .*
>
> *I steal a glance at the faces of those around me: tight lips, wrinkled brows. They stand there with their tails between their legs, with the guilty look of people who know perfectly well that they don't want to do as they must.*
>
> *Not a hand is raised anywhere.*
>
> *The tanks roll, the people keep silent, and the Führer, uncheered, vanishes from the balcony. White-gloved SS men shut the doors, and draw the curtain inside the window.*[52]

The two friends were not the only witnesses to this dismal procession. William Shirer, watching the spectacle unfold, noted in his diary: "Hitler looked grim, then angry, and soon went inside, leaving his troops to parade unreviewed. What I've seen tonight almost rekindles a little faith in the German people. They are dead set against war."[53]

Gisevius also remembered the day: "Never had soldiers been treated so badly in Berlin as they were on that day. In the workers' quarters clenched fists were raised against them; in the center of the city people turned conspicuously away."[54] Paul Schmidt was startled by the "completely apathetic and melancholy behaviour of the Berlin populace."[55]

Hitler withdrew from the balcony and took out his anger on Goebbels: "I can't lead a war with such people." The propaganda minister, who had earlier walked in disguise among the hostile crowd, quickly agreed: "No, my Führer, I saw for myself down there. These people still need enlightenment."[56] With Plan Green about to be launched in three days, Hitler faced a population that, even in a police state, had demonstrated its revulsion against going to war.

SEPTEMBER 27, LONDON, NUMBER 10 DOWNING STREET, 8:00 P.M.

On the same day that Berliners demonstrated their antipathy toward impending war, Londoners resolutely prepared for a war they believed was about to erupt. Open spaces in the city center bristled menacingly with newly installed antiaircraft guns, while government functionaries distributed gas masks to civilians and dug trenches in public parks. Public opinion toward war had shifted dramatically in Britain when the press had reported on Hitler's intransigence at the Bad Godesberg conference. Ordinary Britons believed that Hitler had gone too far. They faced the possibility of war with determination tinged by apprehension. All Britons, from prime minister to chimney sweep, believed that war would bring the roar of German bombers raining death from the skies, bringing to them the same fate that had befallen Basque villagers during the Spanish civil war—horrors tellingly rendered by Picasso in his 1937 painting *Guernica*.

At Downing Street technicians had been at work for an hour, wiring the prime minister's office for his broadcast to the nation. Just minutes before, he had been in a knock-down-drag-out fight with his chief ministers. As Alec Cadogan described it, Chamberlain, Sir Horace Wilson (now returned from Berlin), Halifax, and he had been squabbling over what to tell the Czechs. "Horace drafted a telegram of complete capitulation—telling Czechs to accept Hitler memo. H[alifax] played up against it, and I spoke my mind. Poor P.M. (quite exhausted) said 'I'm wobbling about all over the place' and went in to broadcast."[57] That night the people of Great Britain heard a message that was profoundly confusing.

In contrast to Hitler's belligerence in his radio address at the Sportpalast the night before, the tone of Chamberlain's talk was sorrowful and apologetic. After thanking his countrymen for their letters of support, he turned to the extraordinary war preparations that had been underway in the last twenty-four hours. "How horrible, fantastic, incredible it is that we should be digging trenches and trying on gas masks here because of a quarrel in a faraway country between people of whom we know nothing," Chamberlain said, blithely ignoring the fact that the British Empire had been built on just such wars. "It seems more impossible," he continued, "that a quarrel which has already been settled in principle should be the subject of war," failing to mention that the Czechs and the French did not see it that way.

He addressed the people, urging them not to be alarmed by war preparations, which "do not necessarily mean that we have determined on war or that war is imminent." After these consoling but misleading words, Chamberlain suddenly changed tone:

> However much we may sympathize with a small nation confronted by a big and powerful neighbour, we cannot in all circumstances undertake to involve the whole British Empire in war simply on her account. If we have to fight it must be on larger issues than that. . . . Armed conflict between nations is a nightmare to me, but if I were convinced that any nation had made up its mind to dominate the world by fear of its force, I should believe it must be resisted. Under such domination life for people who believe in liberty would not be worth living.

Thus Chamberlain publicly aired the issue that had dominated cabinet debate, and ministers' private conversations, over the past month. If Hitler was aiming at the domination of Europe, then Britain must try to stop him. But if the Sudeten issue was merely a local quarrel—and the prime minister was convinced that it was—then it need not concern the British Empire.

Alec Cadogan called September 27 "the worst [day] I ever spent."

SEPTEMBER 27, LONDON, NUMBER 10 DOWNING STREET, 9:30 P.M.

The cabinet assembled after the Prime Minister's radio broadcast. The previous day Chamberlain had told the cabinet that, at the urging of Winston Churchill, he had decided to call an emergency meeting of Parliament—its first since July 26—for September 28. The cabinet needed to be brought up to date about recent events, and to decide what approach to take in the Commons the next day.

The prime minister read a telegram from Ambassador Henderson, advising his government to warn the Czechs to accept Hitler's terms or face "the fate of Abyssinia [Ethiopia]," which had been overrun by the Italian army in 1935–36. Then Sir Horace Wilson reported on the failure of his mission to Hitler. During two meetings—the first just hours before Hitler's address the night before, the second that morning—the führer had refused to budge from the position in the Bad Godesberg memorandum. The Czechs would either have to withdraw from the "red areas" marked on the map (most of the Sudetenland), or be overrun by the Wehrmacht. When Wilson had warned of Britain's likely involvement in a war, Hitler "had taken the message very quietly," but in the end had said only, "the Prime Minister [should] do all he could to induce Czechoslovakia to accept his Memorandum."[58] He had given Wilson a letter for Chamberlain reiterating this message.

Chamberlain asked the cabinet if His Majesty's Government should advise the Czech government in a telegram to accept its fate and withdraw beyond the red areas. Duff Cooper said that to do so would be tantamount to urging Czechoslovakia to surrender. He "thought this course was quite unjustified and . . . could not be associated with it," a blunt threat to resign if the cabinet approved sending the telegram. Even

Chamberlain's inner circle began to crack. Sir John Simon expressed grave doubts. None of Chamberlain's supporters came to his rescue.

Halifax said that the suggestion to press the Czechs to accept Hitler's ultimatum "amounted to complete capitulation to Germany." And, "we could not press the Czechoslovakian Government to do what we believed to be wrong." Simon supported Halifax openly. Staring at the bleak reality of a cabinet mutiny, the prime minister yielded: "If that was the general view of his colleagues, he was prepared to leave it at that." The telegram would not be sent.

Now the meeting took on a new tone. The cabinet agreed that when Chamberlain addressed the Commons the next day, he should take the position that if France became involved in a war in defending the Czechs, "we should feel obliged to support them." And, in the most important move that any government could make to prepare the country for war, Chamberlain announced that he had authorized the First Lord of the Admiralty to mobilize the fleet. Chamberlain sat silently in the cabinet room, contemplating the new reality. All his efforts to avoid war had gone awry. His appeasement policy was in tatters, and there seemed to be nothing he could do to salvage it.

When the government's mobilization of the British fleet was announced in Berlin, Ian Colvin noted that it "had more effect than all the words they had hitherto spoken."[59]

SEPTEMBER 27, BERLIN, NUMBER 118 EISENACHSTRASSE, MIDNIGHT

Heinz had assembled his raiding party and sequestered them in several apartments in central Berlin, close to their target on Wilhelmstrasse.[60] Canaris had ordered Groscurth to supply Heinz with carbines and explosives.[61] A declaration of war was expected momentarily.

As they waited for the order to march, Heinz quietly told his men that, whether or not the führer or his SS bodyguard offered resistance, they should provoke an incident and kill Hitler.[62] Heinz surveyed his shock troops of revolution. The commandos eagerly awaited the order from their leader to strike. These determined young men understood the enormity of their task. In Berlin the preparations had been thorough and complete. The fate of the conspiracy would be decided in London.

SEPTEMBER 28, BERLIN, ARMY HEADQUARTERS, VERY EARLY MORNING

Heinz called out his raiding party from the various apartments in which they had been sequestered and assembled them at Army Headquarters. He expected that this would be the day that he would make an end of the National Socialist regime. Heinz handed out the arms, ammunition, and hand grenades that Groscurth had given him. The silence of predawn Berlin was broken by the click-click-click of ammunition being loaded into carbines and automatic weapons. Now, before he could turn his young lions loose on Hitler's Reich Chancellery, he had only to hear word that Brockdorff's Twenty-third Division was on the march from Potsdam.[63]

SEPTEMBER 28, BERLIN, ABWEHR HEADQUARTERS, EARLY MORNING

Late the previous night, by way of Erich Kordt, Oster had received a copy of the letter that Hitler had sent to Chamberlain via Sir Horace Wilson. In the early morning Oster sent Gisevius with a copy of the letter to Halder's office, where Witzleben was waiting. Witzleben read the letter and in turn gave it to Halder. Now they had proof that Hitler was not bluffing and that he was determined on war.

"Witzleben insisted that now it was time to take action. He persuaded Halder to go to see Brauchitsch," Gisevius remembered. "After a while Halder returned to say that he had good news: Brauchitsch was also outraged and would probably take part in the *Putsch.*"[64]

Witzleben immediately reached for the telephone on Halder's desk and called Brauchitsch. He told him that everything was ready, and pleaded with him to issue the order authorizing the coup. Brauchitsch did not say yes or no. Before he would be willing to act he needed to visit the Chancellery. "While Brauchitsch headed for Wilhelmstrasse," Gisevius recalled, "Witzleben rushed back to his military district headquarters," announcing excitedly, "'Gisevius, the time has come!'" Gisevius retraced his steps to Abwehr Headquarters, where Oster remained at his desk prepared to alert Heinz and his commandos to move once Witzleben had given Brockdorff and the Twenty-third Infantry Division its marching orders. Meanwhile, Brauchitsch never got to the Chancellery. On the short walk between his office and Hitler's, the commander in chief learned of unexpected new developments.[65]

SEPTEMBER 28, BERLIN, FOREIGN MINISTRY, MORNING

Erich Kordt had heard from Witzleben that he was prepared to act independently should he receive notification of army mobilization if Brauchitsch hesitated.[66] Just then he received a call from his brother in London. Theo was frantic. He said that it was now a certainty that, in the event of war with Czechoslovakia, "England will fight on the enemy side against Germany." Everyone was awaiting Chamberlain's statement of intentions at the meeting of the House of Commons that afternoon. Theo concluded as forcefully as possible, "If we attack Czechoslovakia we will not have a small war but a big war. Tell this to the responsible persons!"[67]

Just as Erich hung up, Schulenburg knocked and opened his office door. He had news from Oster. "Brauchitsch is apparently ready to participate. I am supposed to find out if the foreign political situation is still the same," he said. "The big war can break out at any time now," Kordt answered him. "We need to act right away before our plot is discovered," Kordt implored Schulenburg, "Don't wait until the afternoon or tomorrow." He assured Schulenburg that there were no special security measures in force at the Reich Chancellery. "I can get you inside the Chancellery. If some of us are there we might be able to open the big double doors behind the guard and clear the way for the raiding party." Schulenburg left to carry the message back to Oster.

September 28, Berlin, Reich Chancellery, Noon

General Jodl called September 28 "the hardest day."[68] Britain and France had taken measures that amounted to partial mobilization. Fritz Wiedemann remembered that September 28 was "the most exciting day I had ever experienced in the Reich Chancellery."[69] Ambassadors, ministers, generals, and rumors were swirling about. Everyone wanted to talk to Hitler, Schmidt wrote. "Whoever happened to be near could get at him, but nobody could actually get a word in. To anyone who wanted, or did not want, to listen, Hitler just made long harangues about his view of the situation."[70]

Gradually, however, Hitler calmed down. "From my corner of the room," Schmidt noted, "I closely watched the actors in this tense battle for peace. I observed from Hitler's reactions how, very gradually, the balance tilted in favour of peace."[71]

The führer was considering the prospect of going to war with the Czechs. Göring advised him against it; Ribbentrop wanted him to invade. Earlier that day Hitler had received a final, desperate letter from Chamberlain, suggesting a five-power conference (Britain, France, Germany, Italy, and Czechoslovakia) to solve the Sudeten dilemma. He had also received a plea from President Roosevelt, urging him not to march. These did not move him.

Suddenly Bernardo Attolico, the Italian ambassador, burst into the room asking to speak to Hitler "on an urgent matter." He ran into Erich Kordt and said in English, "Kordt, I have a personal message from Il Duce. I must see the Führer at once." Attolico assured Hitler that he had Mussolini's support no matter what he decided to do. "The Duce is, however, of the opinion that it would be wise to accept the British proposal, and begs you to refrain from mobilisation." Mussolini proposed a four-power conference—no need for the Czechs—and Hitler was finally swayed. He said quietly, "Tell the Duce that I accept his proposal."[72]

Hitler ordered invitations to be sent to Chamberlain, Daladier, and Mussolini to attend a conference to begin the next day in Munich. A few hours later the führer gave the order for assault troops to be pulled back from their forward positions along the Czech border.

As far as Hitler was concerned, his apparent reversal of strategy indicated only a change in his timetable for war. While he had had great success—and no doubt taken great satisfaction—from bullying the representatives of Britain and France and deceiving them about his true objectives, things had not gone entirely his way.

Plan Green had called for complete surprise. That was no longer possible. The Czechs were ready in their defensive positions, thirty-four divisions strong, waiting for the Wehrmacht's blow to fall. The French had recovered their nerve and now seemed ready to fight. If they did attack with twenty-three divisions in the West, Hitler knew that his nine divisions behind an unfinished West Wall would not hold them out forever. Now even the *British* seemed resigned to fight. Their mobilization of the fleet on the previous day was ominous. What was most disconcerting, however, was that his own people seemed less than eager for war. He had not forgotten the vision of Berliners, silent and sullen as his young warriors paraded along the Wilhelmstrasse, bearing witness to the people's abhorrence of another war.

Hitler knew that Chamberlain was ready to surrender everything, but somehow appeasement had fallen prey to the odd machinations of democracy. At the same time Hitler was confident that at a four-power meeting, with a timorous Chamberlain and a pliant Daladier, he could achieve his goals through bluff and bluster.

At the end of the day Jodl scribbled in his diary, "tension finds relief."[73]

SEPTEMBER 28, LONDON, HOUSE OF COMMONS, 3:00 P.M.

The prime minister was on his feet addressing the first meeting of the House of Commons since July. Members listened attentively as Chamberlain recounted events of the previous two months. The Queen Mother (Queen Mary, widow of King George V, mother of the Duke of Windsor and George VI) and other members of the royal family, clergy, ambassadors, and other notables jammed the Visitors' Galleries. Early that morning Chamberlain had pleaded with Mussolini to intervene. He did not know whether the Duce had been able to convince his fellow dictator to save the peace.

Suddenly Alec Cadogan appeared at the back of the House, carrying two sheets of paper. The note made its way through the tangle of back benchers to Sir John Simon, who was sitting next to Chamberlain. After several failed attempts Simon finally got the prime minister's attention and handed him the missive Cadogan had delivered. Chamberlain stopped speaking and read it. He said to Simon, in a whisper that could be heard throughout the dead-silent House of Commons: "Shall I tell them now?"[74] Simon nodded gravely. Chamberlain turned back to face the House.

> *That is not all. I have something further to say to the House yet. I have now been informed by Herr Hitler that he invites me to meet him at Munich to-morrow morning. He has also invited Signor Mussolini and M. Daladier. Signor Mussolini has accepted and I have no doubt M. Daladier will also accept. I need not say what my answer will be. [An Hon. Member: "Thank God for the Prime Minister!"] We are all patriots, and there can be no hon. Member of this House who did not feel his heart leap that the crisis has been once more postponed to give us once more an opportunity to try what reason and good will and discussion will do to settle a problem which is already within sight of settlement. Mr. Speaker, I cannot say any more. I am sure that the House will be ready to release me now to go and see what I can make of this last effort. Perhaps they may think it will be well, in view of this new development, that this Debate shall stand adjourned for a few days, when perhaps we may meet in happier circumstances.*

The House rose as one—with the notable exceptions of Anthony Eden, Leo Amery, and Harold Nicolson (all anti-appeasement MPs)—

and endorsed the prime minister with wild applause and shouts of encouragement. Harold Macmillan (a young Churchillian MP and future prime minister) ("I stood up with the rest, sharing the general emotion") remembered, "I saw one man silent and seated—with his head sunk on his shoulders, his demeanour depicting something between anger and despair. It was Churchill." But when Chamberlain passed by him, Churchill rose, shook his hand, and said, "Godspeed."[75]

September 28, Berlin, Late Afternoon

Oster and Gisevius waited anxiously. "We could not understand why neither Brauchitsch nor Halder sent word. The minutes passed into hours of unutterable suspense." Meanwhile, at the Foreign Ministry, Schulenburg left Kordt's office and delivered the crushing news to the military conspirators. Gisevius remembered that "the sensational report crashed down upon our heads. The impossible had happened. Chamberlain and Daladier were flying to Munich." Gisevius knew then that "our revolt was done for."[76]

September 29, Berlin, Hans Oster's Apartment

The conspirators awaited news of developments from Munich. They knew that at the conference table in Bavaria their hopes for Germany and for the world were oozing away as documents were discussed and maps were consulted. Hitler would once again emerge triumphant.

With heavy sarcasm, Witzleben expressed the bitterness they all felt. "You see, gentlemen, for this poor foolish nation he is once again our big dearly beloved führer, unique, sent from God, and we, we are a little pile of reactionary and disgruntled officers or politicians who dared to put pebbles in the way of the greatest statesman of all times at the moment of his greatest triumph." Witzleben thought about their legacy. "If we tried to do something now, history, and not just German history, would have nothing else to report about us than that we refused to serve the greatest German when he was the greatest, and the whole world recognized his greatness."[77]

After finishing negotiations with Hitler in the morning, a weary but satisfied prime minister had flown back to England, stopped off at Buckingham Palace to brief the king, and then faced his cabinet colleagues to explain the settlement that he had negotiated at Munich.

Chamberlain's journey from Heston Airport had been a triumphal procession, with crowds cheering wildly as he passed. Chamberlain himself described it: "As I drove from Heston to the Palace [streets] were lined from one end to the other with people of every class, shouting themselves hoarse, leaping on the running board, hanging on the windows, and thrusting their hands into the car to be shaken." At Number 10 the cheering reached a crescendo ("Good old Neville!" "For he's a jolly good fellow!"). Chamberlain had appeared at the window, waved a copy of the agreement that he had just signed with Hitler, and told the adoring crowd, "My good friends, this is the second time in our history that there has come back from Germany to Downing Street, peace with honour. I believe it is peace for our time."[78]

Chamberlain told his Cabinet colleagues that the Munich agreement was "a vast improvement" over the Bad Godesberg memorandum. "It represented an orderly way of carrying out the Franco-British proposals. He thought it was a triumph for diplomacy." But was it? Chamberlain carefully enumerated the ten differences between the two documents, and even an old skeptic like Duff Cooper said that despite his intention to resign, he thought "that the differences between the Godesberg Memorandum and the Munich Agreement were much greater than he had previously recognised."[79]

When calmer heads had an opportunity to analyze the two documents, however, it became clear that Hitler had given away nothing of substance. German troops would still occupy the Sudetenland, now by

October 10 rather than October 1. The question of the Hungarian and Polish minorities remained unresolved. An "International Commission" had been provided for to address technical issues. Plebiscites would be held in the near future, but in areas already under the control of the Wehrmacht.

Why had Chamberlain agreed to such humiliating terms?

Chamberlain's *real* agenda at Munich, which had been his real agenda for the previous ten months, had been the general settlement. He had approached Hitler earlier that morning, after the substance of the agreement had been decided. After small talk designed to put the grumpy führer into a better mood, he had produced a draft of a statement. A weary Hitler, yawning and bored, signed it. The short, three-paragraph statement was vacuous in the extreme. The heart of it affirmed "the desire of our two peoples never to go to war with one another." This toothless smile of a statement was, as Chamberlain assured his listeners, the foundation of "peace for our time."

September 30, London, Foreign Office

Oliver Harvey, Halifax's private secretary, scribbled in his diary that his master was highly skeptical of the Munich settlement. "'I wonder,' is what he says. He has lost all his delusions about Hitler and now regards him as a criminal lunatic. He loathes Nazism." Yet the Munich settlement had been signed, and the Czech Government would "have to be told brutally to agree at once." Harvey could not suppress his bitterness. "I fear," he wrote, "it looks only too much as if we had presented Germany's ultimatum for her."[80]

September 30, On the Road from Munich to Berlin, Evening

Hitler had left the Munich Conference with nearly everything he had demanded. Was he satisfied? "That fellow [Chamberlain] has spoiled my entry into Prague!" Schacht heard him say to his SS entourage on their return to Berlin. Hitler's dream of a conqueror's parade through the streets of the defeated capital city had been punctured by the prime minister's acquiescence to his every wish. Later Hitler said about the Munich settlement, "It was clear to me from the first moment that I could not be satisfied with the Sudeten–German territory. That was only a partial solution."[81]

SEPTEMBER 30, BERLIN, FOREIGN MINISTRY

Walter Hewel, Ribbentrop's liaison officer to Hitler, told Erich Kordt that Hitler had not canceled his war plans; he had just postponed them. Several days later Oster confirmed this, saying that orders had already been given to prepare occupation plans for the remainder of Czechoslovakia.[82]

SEPTEMBER 30, BERLIN, EVENING

Gisevius remembered that as a result of the outcome of the Munich conference the conspirators "were extremely depressed." They were convinced that the Munich agreement would not secure peace for Europe; rather "it was a signal for a world war." Gisevius, Schacht, and Goerdeler discussed their alternatives. Should they emigrate? Goerdeler said that "to be able to continue our political work at all in Germany in the future there is only one possibility, and that is to employ the methods of Talleyrand." Gisevius noted with resignation, "We decided to persevere."[83]

The Czech crisis had created a unique opportunity for Germans to throw off the oppressive rule of Nazism. Last-minute British capitulation at Munich had crushed the conspirators' plans. They saw no alternative but to return to work, stay alert, and await the next chance for action.

OCTOBER 3, BERLIN, WITZLEBEN'S HOUSE

A few days after the Munich conference, Gisevius, Schacht, and Oster sat around the grand fireplace in Witzleben's house "and tossed our lovely plans and projects into the fire," according to Gisevius. The security risk of doing otherwise was just too great. "We spent the rest of the evening meditating, not on Hitler's triumph, but on the calamity that had befallen Europe."[84]

OCTOBER 3–5, LONDON, HOUSE OF COMMONS

The House of Commons debated the Munich agreement for three days. Duff Cooper, as promised, resigned from the cabinet, and was accorded the opportunity to speak first. In contrast to the prime minister, who had addressed the Nazis in the language of "sweet reasonableness," Duff Cooper would have used "the language of the mailed fist." But what of the guarantee that Britain had proffered to the Czechs? "We have taken away the defences of Czechoslovakia in the same breath as we have guaranteed them, as though you were to deal a man a mortal blow and at the same time insure his life." He concluded: "I have ruined, perhaps, my political career. But that is little matter. I have retained something which is to me of great importance. I can still walk about the world with my head erect."[85]

The House listened respectfully to Duff Cooper's departing speech. Other speakers were not so gently treated. Labour MPs and some Conservatives, like Anthony Eden, attacked the agreement, and in turn drew cries of "Shame!" or "Peace!" from the Conservative back benches. For nearly three days Churchill sat silent and brooding. Not until 5:10 P.M. on October 5 did the great dissenter rise to speak. He adopted his customary stance, facing the House with thumbs hooked into his waistcoat pockets, feet firmly planted beneath his considerable bulk. He didn't pull his punches. At the beginning of his speech he said, "We have sustained a total and unmitigated defeat, and . . . France has suffered even more than we have." "Nonsense!" shouted his rival Conservative MP Lady Nancy Astor, who had been born in America but became the social center of the "Cliveden Set," which strongly supported appeasement.

Churchill summed up the negotiations from Berchtesgaden to Bad Godesberg to Munich: "They will be very simply epitomized, if the House will permit me to vary the metaphor. One pound was demanded at the pistol's point. When it was given, two pounds were demanded at the pistol's point. Finally, the dictator consented to take one pound, seventeen shillings and sixpence and the rest in promises of good will for the future." Churchill reminded the House that there had been alternatives that the government had spurned. Britain could have repaired ties with her allies—France, the United States, perhaps even Russia—and constructed an alliance "of Powers great and small, ready to stand firm upon the font of law and for the ordered remedy of grievances."

Churchill predicted a gloomy end for the Czechs. "All is over. Silent, mournful, abandoned, broken, Czechoslovakia recedes into darkness," he said prophetically. "She has suffered in every respect by her association with the Western democracies."

On the subject of Britain and whether it would ever recover the moral fiber that had been sacrificed for peace, he said, "And do not suppose that this is the end. This is only the beginning of the reckoning. This is only the first sip—the first foretaste of a bitter cup which will be proffered to us year by year—unless by a supreme recovery of our moral health and martial vigour, we arise again and take our stand for freedom, as in the olden time."

Chamberlain carried the House on the motion to approve the Munich agreement, 366–144.

MARCH 15, 1939, PRAGUE

Precisely one year after the Nazi takeover of Austria, the citizens of Prague awoke to the unmistakable rumble of tank treads on the cobblestone streets of their ancient and beautiful city. The Wehrmacht had arrived. The Slovak lands were split off to become a German puppet state. Bohemia and Moravia were reduced to a "protectorate" of the Third Reich. The dismemberment of Czechoslovakia was complete, and the Nazi domination of Europe had just begun.

Erich Kordt wrote, "The Munich Conference prevented the coup d'état in Berlin." Gisevius was even more blunt: "Chamberlain saved Hitler."[86]

Epilogue

On the night of November 7, 1939—two months after Germany's successful campaign against Poland, which ushered in World War II—Franz Maria Liedig drove Hans Oster to the Berlin apartment of Colonel Gijsbertus Sas, the Dutch military attaché in Berlin and Oster's good friend since 1936. During the drive Oster was very quiet. Liedig waited in the car while Oster entered the apartment to talk to Sas. When Oster returned, his garbled sentences were barely comprehensible. But Liedig soon understood that his friend was deeply troubled, and that he had "crossed the Rubicon." Oster said quietly, "There is no way back for me." Liedig asked him what he meant, and Oster replied, "It is much easier to take a pistol and kill somebody; it is much easier to run into a burst of machine-gun fire than it is to do what I have done."[1]

Liedig later discovered the source of his friend's discomfort. Oster had just revealed to Sas the entire German plan for the invasion of Western Europe, then scheduled for November 12. Hitler kept postponing the attack, but Oster would continue to keep Sas fully informed, for the last time on May 9, 1940, less than twelve hours before the attack actually occurred. On that night the two men went out for dinner—which Sas called "funereal"—after which Oster stopped briefly at Army Headquarters for a few minutes and then returned to the car. "My dear friend, now it is really all over. . . . The pig [Hitler] has gone off to the western front, now it is definitely over. I hope that we shall meet again after this war."[2] Sas dutifully passed on the invasion plans to his superiors, who completely ignored Oster's precious gift. When the attack came on May 10, at exactly the time and in exactly the form that Oster had predicted, the Dutch, Belgian, and French commanders were caught quite unprepared.[3] Oster never had any ethical doubts about his decision to reveal the invasion plans, an action that, under ordinary circumstances, would be treasonous. He told Sas, "One might say that I am a traitor (*Landesverräter*), but in reality I am not; I consider myself a better German than all those who run after Hitler. It is my plan and my duty to free Germany, and at the same time the world, of this plague."[4]

In September 1938, the failure of the Oster conspiracy did not bring an end to the resistance activities of this fierce opponent of Nazism. Indeed, like most of the 1938 conspirators, Oster continued to try to undermine the Third Reich. He worked to prevent the outbreak of war right up until the German invasion of Poland on September 1, 1939. By that time, however, the conspirators' arguments were falling on deaf ears. Germany was far better prepared for war than it had been a year earlier. The flow of men and armaments had continued to swell the Wehrmacht. The Czechs had been eliminated as a potential enemy, and their gold and arms had been added to the treasury and arsenal of the Third Reich. Britain and France were revealed, as Hitler had predicted, to be toothless tigers. In August 1939 the notorious Nazi-Soviet Pact had transformed the Soviet Union from foe into ally. German officers looked at the possibility of war with much less trepidation than they had in September 1938, and the enthusiasm of the generals for a coup waned.

Late in the afternoon of August 31, 1939, with the outbreak of war just hours away, Oster summoned Gisevius to Abwehr Headquarters. Admiral Canaris, in the company of other officers, spotted Gisevius. After his colleagues left the room, Canaris grasped him by the arm and asked, "Well, what do you say now?" Without waiting for Gisevius's reply, Canaris himself supplied the answer: "This means the end of Germany."[5]

The next day Germany launched its attack on Poland. The Poles, overwhelmed by the dive-bombers and panzers of the Wehrmacht, were defeated in a matter of weeks. After Poland the guns fell silent—ushering in a period aptly named the "phony war." Europeans waited apprehensively for Hitler to make a move. Would he stop with Poland? the British and French wondered. Would he turn his guns on them next? Or would he continue to press eastward, toward the Soviet Union? In that case their countries might be spared, at least temporarily, and they would have more time to rearm.

Britain and France were technically at war with Germany, but neither side was shooting, and the Dutch and Belgians were still neutral. Both the British government and the conspirators sensed that the sudden lull in the fighting created a unique opportunity. If Germany and Britain could reach a negotiated settlement before the two sides unleashed their heavy weapons against each other, peace might still be restored, particularly if Oster, Beck (now returned to an active role in the conspiracy), and their collaborators could overthrow the Nazi regime and replace it with a less belligerent government.

In the autumn of 1939, with the agreement of Canaris and Beck, Oster sent for Josef Müller, a Catholic lawyer from Bavaria. Müller remembered the dramatic meeting at Abwehr Headquarters: "I was received by Colonel Oster, who conveyed General Beck's request that I travel to Rome and explore whether Pope Pius XII was willing to contact the British government and learn about their willingness to negotiate a ceasefire followed by peace negotiations."[6] The only way for Müller to get to Rome was for Canaris to authorize his travel as an Abwehr operative. "Oster then stated that I was now in the center of German resistance and that they needed me to help them to restore peace, which had been blatantly endangered by Hitler." The next day

Canaris himself met with Müller, telling him that Hitler's "waging of war was not only criminal but also unprofessional." Müller agreed to go and left shortly thereafter. With the pope acting as a reluctant middleman, a tentative connection was made. The British by now were willing to listen and to take the plotters seriously.[7]

Even Halifax was willing to risk contact with the German conspirators. He began indirect talks with them through Ulrich von Hassell, a prominent anti-Nazi and a former German ambassador to Italy. Hassell's British contact was J. Lonsdale Bryans, whom Halifax knew socially. Halifax authorized Bryans to encourage the conspirators to overthrow Hitler, and to give them assurances that Britain would look favorably on a new German government. Bryans met with Hassell forty times from the autumn of 1939 through the spring of 1940. Halifax continued to have a wait-and-see attitude toward the conspirators. Talks with Hassell, he said, "can do no harm, and may do a lot of good."[8]

While these contacts were under way in Rome, Weizsäcker posted Theo Kordt to Bern in neutral Switzerland after the outbreak of war. His secret mission was to maintain his British connections and to explore possibilities for peace. Between October 1939 and February 1940, Theo Kordt met clandestinely several times with his old friend Philip Conwell-Evans.[9] Kordt remembered that "the opposition was encouraged by the clear formulation from the British side in October 1939 that 'the removal of Hitler would mean the deaths of thousands instead of millions.' "[10]

There is piquant irony in the sudden willingness of the British government to undertake negotiations with the very group of Germans they had rejected a year earlier, and whose predictions about Hitler's behavior had been so precisely borne out by recent events. In these resumed negotiations, the goals of both sides were clear. The British wanted a peaceful solution to the war, preferably resulting in a new government, German disarmament, and a more effective regime. The conspirators wanted a promise from the British that if they could overthrow the government, perhaps precipitating a civil war in Germany, that the British would not take advantage of the country's temporary weakness by launching a military offensive. In the spring of 1940, His Majesty's Government gave the conspirators this assurance.[11]

Meanwhile, after his quick defeat of the Poles, Hitler sent out a peace overture to the British on October 6, 1939, which was rejected. He decided to attack Western Europe immediately. With the renewed prospect of war facing the Wehrmacht's generals, this time against Britain, France, Belgium, and the Netherlands, the old doubts returned and Oster and Beck were able to reconstruct the conspiracy of September 1938.

The conspirators resurrected their plans of the previous year, making changes to accommodate new circumstances, such as Witzleben's transfer out of Berlin. The key to this renewed plot was General Hoepner, now in command of a panzer army consisting of seven divisions, headquartered in Düsseldorf. Erich Kordt claimed, "The plan was to keep two armoured divisions of Hoepfner [sic] Gruppe in the vicinity of Berlin to use them to surround the Reich Chancellery on the day of issuance of orders for the Western Offensive, and to occupy strategic positions . . . in the rest of Germany." Kordt remembered that the generals lost their nerve. "Hitler apparently suspected conspiracy. I was informed that he intended to take drastic action against what he called 'the decadent spirit' of Zossen (the new Army HQ)."[12]

When the generals who had access to Hitler showed themselves to be unwilling assassins (including Halder, who for several weeks carried a loaded pistol to his meetings with Hitler but could not bring himself to pull the trigger), Erich Kordt agreed to kill Hitler himself. "All I need is a bomb," he told Oster. "You will have the bomb by November 11," was the reply.[13] Kordt was regularly in Hitler's presence in the Reich Chancellery as a member of Ribbentrop's entourage, and had become such a routine fixture that he was no longer subject to searches or identity checks.[14] He intended to strap the bomb to his body and detonate it when he came close to Hitler. The scheduled assassination date was November 11, 1939, one day before the planned invasion of Western Europe. However, a failed bombing attempt on Hitler's life on November 8 by a "lone wolf," Georg Elser, made it impossible for Oster to procure the necessary explosives and led to a tightening of security around the führer. Changing weather conditions forced Hitler to postpone the invasion several times, and Halder and Brauchitsch showed themselves irresolute.[15] Despite Oster's best efforts, the conspiracy plans once again faded away.

While the conspirators continued to scheme, it was not until the fortunes of war had turned against Germany that new opportunities arose. By February 1943 the Germans had suffered a monumental defeat at Stalingrad, where one of Oster's sons was killed in action.[16] With the German defeat on the Eastern Front, the Allied successes in North Africa, the German defeat in the Battle of the Atlantic, and the first bombing raids on German cities, it was clear that the tide had turned and strategic defeat was likely.

In March 1943 Oster and his friend and Abwehr colleague Hans von Dohnanyi were instrumental in an attempt to kill Hitler by some officers of Army Group Centre, then deep inside the Soviet Union. The Abwehr contingent had actually carried captured British explosives—which had technical advantages over German explosives—to Smolensk, where they were to be planted in the aircraft that Hitler would take from the Russian front to his headquarters in East Prussia on March 13. Despite extremely tight security, the conspirators managed to plant the bomb and the airplane took off. They waited nervously for several hours for word of the expected "accident." Once Hitler had been killed, Oster and others back in Berlin would seize control of the capital. Alas, the bomb failed to detonate, perhaps because of the cold temperature.[17] Still the conspirators did not give up. Another attempt later that month in Berlin—using the same package of British explosives—failed when Hitler changed his preset plan for a leisurely review of a military exhibit where he was to be killed by a suicide bomber. By luck and by instinct Hitler had avoided the conspirators' traps.

Hans von Dohnanyi had joined the Abwehr in 1939, and had worked continuously in resistance activities with Oster. Both were concerned about Hitler's increasingly deadly war against the Jews. On November 9, 1938, the assassination of a German diplomat by a Jew in Paris provided the excuse for Kristallnacht, the widespread Nazi vandalism against Jewish-owned property (the broken glass of Jewish shops giving the night its doleful name), and the first roundup of Jews for transportation to concentration camps. Kristallnacht shocked those Jews who had believed that Nazi anti-Semitism would blow over into realizing that they had to leave. Despite the hardships and difficulties they faced in leaving, another 150,000—equal to the total emigration in the previous five years—left Germany within months.[18] It was only in 1941, however—after appallingly brutal roundups, concluding with mass fusillades of Jewish men, women, and children in open fields, proved too public, too slow, and too harrowing for the executioners— that the SS began to create death camps. Here Jews from all over occupied Europe would meet their hideous end in chambers specifically designed to kill with the poison gas Zyklon-B.

Beginning in 1943, well aware of the Jews' desperate plight, Oster and Dohnanyi tried to help a few of the remaining German Jews to make their escape. They mananged to provide fourteen Jews with the necessary funds and documentation as "Abwehr officers" to escape to neutral Switzerland. The Gestapo caught wind of their financial irregularities, which involved substantial amounts of money. At first suspecting only corrupt profiteering, the Gestapo arrested Dohnanyi on April 5, 1943 and charged him with illegal foreign currency transactions. After being interrogated by the Gestapo and tried, Dohnanyi was incarcerated in Sachsenhausen concentration camp.

Oster's detection and house arrest in 1943 resulted from the same currency manipulations that had snared Dohnanyi.[19] As he had so many times in the past, Oster had been able to cover his own tracks. However, in his attempt to protect Dohnanyi, he was tripped up. Oster was tried on September 16, 1943, for attempting to hide and destroy evidence of Dohnanyi's currency dealings and for aiding associates, particularly Pastor Dietrich Bonhoeffer, Dohnanyi's brother-in-law, evade military service by "illegal transfers" to the Abwehr.

Ernst Kaltenbrunner, who had replaced Reinhard Heydrich as head of the Gestapo, saw the charges against the Abwehr officer as a way of getting rid of a rival and a reluctant servant of the regime. Oster was discharged from the Abwehr and placed under house arrest.[20] He remained limited in his movements and unable to engage in the preparations leading up to the July 1944 plot against Hitler.[21]

The *Reichsdepositenkasse* affair, which had led to the arrest of Dohnanyi and Oster, had also implicated Canaris himself. Although he survived for a few months, he was politically wounded. When Hitler became disenchanted with the admiral for other reasons, Canaris's rivals Himmler and Ribbentrop moved in for the kill. In February 1944 Canaris was removed from his post and put under house arrest while the entire Abwehr was dissolved and its parts dispersed to other Nazi agencies.[22]

In July 1944, after the bomb that Colonel Claus Schenk von Stauffenberg left in a briefcase in the führer's headquarters in East Prussia failed to kill Hitler, the Gestapo launched a massive hunt, not only for the immediate perpetrators, but for all resisters. Among the approximately seven hundred people arrested, several mentioned Oster as a leading resister, even though he had not participated in this particular assassination attempt. Oster was taken to Gestapo headquarters for interrogation. He fought for his life as the Third Reich collapsed, keeping his interrogators engaged but at bay in a deadly battle of wits. He was nearly successful. At the end of March 1945, with an Allied victory only weeks away, Oster was still alive.

Most of Oster's military collaborators were arrested in the wake of Stauffenberg's failed bomb plot. Generals Erwin von Witzleben, Erich Hoepner, and Ludwig Beck had joined the Stauffenberg plot. When it failed all three were killed—Beck on the night of the plot's collapse, and Witzleben and Hoepner several weeks later, when they were tried before the notorious "People's Court" and hanged by a piano wire suspended from a meat hook.

Colonel General Franz Halder continued to work with Oster, particularly in the abortive coup attempt of October–November 1939. He served the Third Reich as chief of the General Staff until 1942, when a series of conflicts with Hitler, including a disagreement over the strategy that led to the Battle of Stalingrad, forced his resignation. Although Halder was not involved in the Stauffenberg plot he was arrested and interrogated. He was spared execution but sent to a concentration camp, where he survived the war.

Hans Oster's principal allies in the Foreign Ministry survived the war because of fortuitous assignments. In 1941 Erich Kordt was posted to the German Embassy in Japan, and spent the remaining war years in Tokyo and then in Japanese-occupied Shanghai. His cousin, Susanne Simonis, went with him and served as the bachelor Erich's hostess in both postings. After the establishment of the Federal Republic of Germany (West Germany), he became an adviser to the Foreign Ministry in the new Christian Democratic government. In 1951 Kordt became a lecturer at the University of Cologne, and in the later 1950s served as a ministerial adviser to the state government of North Rhine–Westphalia in the heavily Catholic Rhineland area. After the war Susanne Simonis became the first woman to join the West German Foreign Service. She was posted to London, to Scotland, and then to Bonn, where, as director of the Personnel Department, she helped recruit other women to the Foreign Service. She completed her diplomatic career as the German consul general in Vancouver, Canada, from 1965 through 1969.[23] Theo Kordt remained in Bern for the duration of the war. After the war he was director of the regional department in the Federal Office of West Germany, 1950–53, and was West German ambassador to Greece from 1953 to 1958.

Ernst von Weizsäcker continued as state secretary until 1943, when he persuaded Ribbentrop to appoint him ambassador to the Vatican, where he thought that he might escape the carnage in Germany and perhaps further the cause of peace. For his role of preparing Germany for war, he was tried and convicted by the Americans as a war criminal in 1947, and was imprisoned until 1950.[24]

A few of Oster's fellow conspirators went underground. Hans-Bernd Gisevius drew on his contacts with the American intelligence organization, the Office of Strategic Services (OSS), to make his way into Switzerland in 1945. Friedrich Wilhelm Heinz and Oster remained confidants and colleagues up until Oster's arrest in 1943. When the Gestapo spread its dragnet in 1944, Heinz disappeared. Using his old contacts, he eluded capture and survived the war inside Germany.[25] Carl Goerdeler was not so fortunate. After the failure of the Stauffenberg plot, the ebullient emissary and inveterate planner went into hiding. He was arrested on August 12 when he foolishly took a trip to his hometown to visit his parents' grave, despite a reward of a million marks on his head. He was recognized at a nearby inn and betrayed by an old family acquaintance. Goerdeler was executed on February 2, 1945.

Hjalmar Schacht, the leading civilian among the 1938 plotters, resigned as president of the Reichsbank in 1939, but at Hitler's insistence retained the meaningless title of minister without portfolio until 1943. He was arrested in the wake of the Stauffenberg plot and spent the remainder of the war in concentration camps. At the end of the war he was arrested by the Allies and tried as a war criminal at Nuremberg. Schacht was one of only three defendants to be acquitted of all charges.

Across the North Sea, the fates of Oster's would-be allies Chamberlain, Churchill, and Halifax were intertwined through 1940. When war broke out Chamberlain was forced to name the bellicose Churchill to the cabinet. But Britain was unable to engage the Nazis successfully, and in May 1940 the House of Commons called Chamberlain to account for his flaccid prosecution of the war. He was forced to step down and replaced by Churchill. Under his leadership Britain stood alone against the Nazis for twelve months—from June 1940 through June 1941—until the German attack on the Soviet Union (June 22, 1941) and the Japanese attack on Pearl Harbor (December 7, 1941) brought about the "Grand Alliance" that Churchill had so passionately advocated in 1938. Churchill waged war with vigor and imagination; he is rightly regarded as one of the greatest prime ministers in British history.

The events of May 1940 were the defining moments in Halifax's professional career. On May 9, when it became clear that Chamberlain would have to go, the prime ministership was his for the taking. But Halifax declined, publicly because his title would preclude his leading the government in the House of Commons, but privately because he knew that he lacked the decisiveness and passion to see the country through a long war.

Halifax also played a crucial role as an advocate for a negotiated peace with Hitler in late May, as France was crumbling and the remnants of the British Expeditionary Force struggled to escape from Dunkirk. Churchill's position in the cabinet was not strong enough to allow him to reject Halifax's plan out of hand, but he sensed that Hitler would not make peace until he had tried and failed to invade Britain.[26] The debates in the War Cabinet on May 27 and 28 were acrimonious, with Churchill and Halifax the leading antagonists. Eventually Churchill prevailed, and his judgment was vindicated by the Battle of Britain and Hitler's abandonment of his plans to invade England in the autumn of 1940. Halifax's final service to his country was as ambassador to the United States from 1941 through 1946.

Oster's Nazi enemies perished. Heydrich was assassinated by Czech partisans in 1942. At the end of April 1945, as Russian armies closed to within a few hundred yards of the Reich Chancellery, Hitler and Goebbels committed suicide in the führer's Berlin bunker. Himmler attempted to escape, disguised as an enlisted man. When he was detected and arrested by two British officers on May 23, 1945, he swallowed a poison capsule. Göring, facing a death sentence after his conviction at the Nuremberg Trials, took his own life with a cyanide pill on October 14, 1946. Two hours later Generals Jodl and Keitel, along with Joachim von Ribbentrop, having been found guilty by the Allied tribunal, were hanged.

Hans Oster, Wilhelm Canaris, Hans von Dohnanyi, and Pastor Dietrich Bonhoeffer (who had begun working actively with Oster and Dohnanyi in 1939) were all victims of one of Hitler's final acts of vengeance. In the spring of 1945, with Allied armies on German soil, the Gestapo uncovered documents at army headquarters, including Canaris's diary, which revealed the extent and longevity of military resistance to the regime. Hitler, beside himself with rage at the discovery that they had conspired against him since before the war, ordered the "merciless destruction" (*unbarmherzige Vernichtung*) of all remaining resisters.[27] On April 9, after a summary trial at Flossenburg concentration camp, Oster and his colleagues were slowly hanged to death.[28] In the silence of the early morning, as they mounted the gallows, they might have heard the guns of the approaching American armies.

Cataclysmic changes in history are often the result of seemingly minor decisions taken by individuals. The events of September 1938 are a dramatic example. Suppose that Halifax had carried out his cabinet mutiny against Chamberlain more vigorously in the last days of September, forcing him to abandon appeasement for a more confrontational policy toward Nazism. Or suppose that Hitler had not yielded to Mussolini's desperate pleas to settle the Sudeten crisis at the conference table, and had carried through with his intention to invade the Sudetenland on October 1, 1938. The declaration of war against Czechoslovakia would have triggered the coup that had been so meticulously prepared by Oster and his fellow conspirators. If Hitler had been killed and the Nazis deposed, the coup would have brought to power men dedicated to restoring moral order to Germany and peace to Europe. World War II would never have taken place; fifty million people would not have lost their lives; and the shape of the twentieth century would have been vastly different.

Glossary of German Terms

Abwehr: German armed forces (army, navy, air) intelligence and counterintelligence staff, under Admiral Wilhelm Canaris, whose chief of staff, Hans Oster, made that organization a base for military resistance to Hitler.

Alte Kämpfer: Old fighters: those members of the Nazi Party in the early days, particularly before it was temporarily banned after the Hitler putsch of 1923.

Freikorps: Free corps; right-wing paramilitary forces, composed mainly of monarchist or fascist ex-soldiers who remained under arms despite the limits imposed by the Treaty of Versailles. First used to suppress the Spartakus uprising in Berlin in late 1918, they continued as an antirepublican domestic force, as well as in former German territory in the Baltic states, Poland, and the Ukraine.

Generalität: Corps of Wehrmacht generals.

Gestapo: Geheime Staatspolizei; literally, Secret State Police, the political police already existing in Prussia during the Weimar Republic, which under the Third Reich was amalgamated in 1939–40 with the Nazi Party Security Service and most of its personnel made members of the General SS (Allgemeine SS), as were the uniformed regular police, which after 1936 was headed by Heinrich Himmler in his dual capacity as Reich leader of the SS and chief of the German Police.

Kriegsakademie: War College.

Kriegsmarine: Navy.

Kripo: Kriminalpolizei (Criminal Police), headed by Arthur Nebe, one of Oster's informants.

Lebensraum: Living space—the vast areas in eastern Europe that would be opened to Aryan colonists after the Jewish inhabitants had been exterminated and the Slavic inhabitants reduced to slavery.

Luftwaffe: Air force.

OKH: Oberkommando des Heeres; Army High Command.

OKW: Oberkommando der Wehrmacht; High Command of the German Armed Forces, including the navy (Oberkommando der Marine: OKM), army (OKH), and air force (OKL). From the moment of its creation to replace the War Ministry in 1938, Hitler was the commander in chief of the OKW.

Orpo: Ordnungspolizei; uniformed police.

Reichswehr: German Defense Forces; limited to one hundred thousand troops by the Treaty of Versailles, the Reichswehr, made up largely of former imperial officers and noncommissioned officers, at best remained "nonpolitical" and neutral toward the Weimar Republic during its first years under an SPD president, but after 1925 accepted, if not the republic, its elected President, Field Marshal Paul von Hindenburg, as a surrogate kaiser. By 1935 the Reichswehr was much expanded and renamed Wehrmacht.

SA: Sturmabteilung; storm troopers, brown-shirted street brawlers of the pre-Nazi era who, after the decimation of their leaders in the Röhm Putsch (the so-called Night of the Long Knives) of June 30, 1934, were eclipsed by the black-shirted SS, until then a small elite group within the SA.

SS: Schutzstaffel, literally, defense staff, black-shirted elite formation of the Nazi Party, which under Heinrich Himmler burgeoned into the sources of personnel for political police (Gestapo), security forces (SD), the *Totenkopfverbände*, or Death's Head Units, which manned the concentration camps, and armed SS (Waffen SS) combat units.

Stahlhelm: Another name for the Frontkämpferbund, World War I veterans' organization, enlarged to include a younger generation in subsequent years; most of its members were incorporated into the SA Reserve after the Röhm Putsch of June 1934.

Waffen SS: Armed SS, combat military units of the SS, which during the war grew from a few regiments to many divisions and corps, as well as the command staff of an army group.

Wehrkreise: military districts (twelve in Germany in 1938) of reserve or home army, providing regional bases for recruitment and training.

Wehrmacht: German armed forces, evolved from the Reichswehr and renamed after Hitler repudiated the military restrictions of the Treaty of Versailles in 1935.

Army Rank Structure for General Officers

GERMAN	AMERICAN
Major General	Brigadier General
Lieutenant General	Major General
General of Artillery	Lieutenant General
or	
Infantry	
Colonel General	(Four-Star) General
Field Marshal	General of the Army

Notes

Preface

1. J. W. Wheeler-Bennett, in his highly regarded history of the German army, concludes that "there is no evidence but the flimsiest assertion that, had Mr. Chamberlain never gone to Berchtesgaden or to Godesberg or to Munich, the conspirators would have been sufficiently prepared or resolute to strike." *The Nemesis of Power: The German Army in Politics, 1918–1945* (London, 1953), p. 424. William Shirer, though less vituperative than Wheeler-Bennett, is equally dismissive of the conspirators' chances for success: "Though much would later be written about the German 'resistance' movement, it remained from the beginning to the end a small and feeble thing, led, to be sure, by a handful of courageous and decent men, but lacking followers. . . . Moreover, how could a tiny group—or even a large group, had there been one—rise up in revolt against the machine guns, the tanks, the flame throwers of the SS?" *The Rise and Fall of the Third Reich* (London, 1960), p. 455. He is also skeptical about the claims by a surviving conspirator, Hans-Bernd Gisevius, that, because of his surrender at Munich, "Chamberlain saved Hitler." *To the Bitter End: An Insider's Account of the Plot to Kill Hitler, 1933–1944* (1947;

reprint, New York, 1998), p. 326. Shirer responds: "Did he? Or was this merely an excuse of the German civilians and generals for their failure to act?" (p. 503). See also Gerhard Ritter, *Carl Goerdeler und die deutsche Widerstandsbewegung* (Stuttgart, 1956). Published in English as *The German Resistance: Carl Goerdeler's Struggle Against Tyranny*, translated by R. T. Clark (London, 1958). Hans Rothfels, *Die deutsche Opposition gegen Hitler* (Frankfurt and Hamburg, 1947). Published in English as *The German Opposition to Hitler* (Hinsdale, Ill., 1948). In the late 1960s there was a spirited debate in the pages of *Encounter* over the issue of the British response—or lack of it— to overtures by the German conspirators. Christopher Sykes, "Heroes & Suspects: The German Resistance in Perspective," *Encounter*, December 1968, pp. 39–47; David Astor, "Why the Revolt Against Hitler Was Ignored," *Encounter*, June 1969, pp. 3–13; and a series of letters, "David Astor & the German Opposition," *Encounter*, September 1969, pp. 89–96.

2. Joachim Fest, *Plotting Hitler's Death: The Story of the German Resistance* (New York, 1996), pp. 325, 350. Peter Hoffmann, *The History of the German Resistance, 1933–1945* (1977; reprint Montreal and Kingston, Canada, 1996); first published in German in 1969 as *Widerstand, Staatsstreich, Attentat.* Like all historians of the German resistance, I owe a great deal to Hoffmann's definitive research. Another recent book about a special aspect of resistance history is Klemens von Klemperer, *German Resistance Against Hitler: The Search for Allies Abroad 1938–1945* (Oxford, England, 1994). Klaus-Jürgen Müller has written an overview of resistance historiography in "The Military Opposition to Hitler: The Problem of Interpretation and Analysis," in his *The Army, Politics, and Society in Germany, 1933–45: Studies in the Army's Relation to Nazism* (Manchester, England, 1987), pp. 100–122.

3. See Remedio von Thun-Hohenstein, "Widerstand und Landesverrat am Beispiel des Generals Hans Oster," in *Der Widerstand gegen den National-Sozialismus*, edited by Jürgen Schmädeke (Munich, 1994).

4. Theodore Hamerow, *On the Road to the Wolf's Lair: German Resistance to Hitler* (Cambridge, Mass., 1997), pp. 242–43. See Hoffmann's review of Hamerow in *Historische Zeitschrift* 271 (2000), p. 817.

5. Gisevius, *To the Bitter End*; Fabian von Schlabrendorff, *Offiziere gegen Hitler* (Zurich, 1946). Republished in English as *The Secret War Against Hitler*, translated by Hilda Simon (New York, 1965); and Kordt, *Nicht aus den Akten.* Both Schlabrendorff's and Gisevius's books were published with the encouragement and assistance of Allen Dulles of the OSS. Gisevius, a direct participant in the events, as an author was given to occasional exaggeration. Schlabrendorff, although familiar with the

main conspirators of 1938, was not a direct participant in the events. There are also a few memoirs, published in German, by resisters who were peripheral to the 1938 conspiracy: Josef Müller, *Bis zur Letzten Konsequenz: Ein Leben für Frieden und Freiheit* (Munich, 1976); Hans Speidel, *Aus unserer Zeit: Erinnerungen* (Berlin, 1977); Rudolf-Christoph von Gersdorff, *Soldat im Untergang* (Frankfurt, 1977); Peter Sauerbruch, "Bericht eines ehemaligen Generalstabsoffiziers über seine Motive zur Beteilung am militärischen Widerstand," in *Aufstand des Gewissens: Militärischer Widerstand gegen Hitler und das NS-Regime—Katalog zur Wanderausstellung des Militärgeschichtlichen Forschungsamtes* (Herford, 1984), pp. 421–438; Otto John, *"Falsch und zu Spät": Der 20. Juli 1944* (Munich, 1984); Philipp von Boeselager, *Der Widerstand in der Heeresgruppe Mitte, Beiträge zum Widerstand 1933–1945*, no. 40 (Berlin: Gedenkstätte Deutscher Widerstand, 1990). Still another German-language source on aspects of the conspiracy is the report of Heinrich Müller, chief of the Gestapo, on his investigation of the origins of the 1944 Stauffenberg plot: *Spiegelbild einer Verschwörung: Die Kaltenbrunner-Berichte an Bormann und Hitler über Attentat vom 20. Juli 1944—Geheime Dokumente aus dem ehemaligen Reichssicherheitshauptamt* (Stuttgart, 1961), hereafter cited as *Spiegelbild*. Müller forwarded the reports to his superior, Ernst Kaltenbrunner, who was head of the Reich Security Central Office (RSHA). Kaltenbrunner signed them and sent them on to Martin Bormann, head of the Nazi Party Reich Chancellery and secretary to the führer. The reports are available on (U.S.) National Archives microfilm T4, Rolls 19–21.

6. Harold C. Deutsch, *Hitler and His Generals: The Hidden Crisis, January–June, 1938* (Minneapolis, 1974), and *The Conspiracy Against Hitler in the Twilight War* [1939–1940] (Minneapolis, 1968). Deutsch also published two short articles about the 1938 conspiracy: "German Soldiers in the 1938 Munich Crisis," in Francis R. Nicosia and Lawrence D. Stokes, eds., *Non-Conformity, Opposition and Resistance in the Third Reich* (London, 1990), pp. 305–21, and "Dress Rehearsal Crisis 1938," in Harold C. Deutsch and Dennis E. Showalter, eds., *What If? Alternatives of WWII* (Chicago, 1997), pp. 9–25. Deutsch had had the rare experience for an American scholar of having witnessed with alarm the atmosphere of the Third Reich while doing research there in the 1930s.

7. The Deutsch Papers (U.S. Army War College, Carlisle, Pennsylvania), though still uncataloged, are available to researchers. I found thirty-six files relevant to the Oster conspiracy.

Prologue

1. Erich Kordt, *Nicht aus den Akten: Die Wilhelmstrasse in Frieden und Krieg* (Stuttgart, 1950), pp. 249–52.

Chapter 1: The Führer Makes War on His Army

1. *Spiegelbild*, pp. 32–3, 451. See also Remedio von Thun-Hohenstein, *Der Verschwörer: General Oster und die Militäropposition* (Berlin, 1982), pp. 45–47. By the 1930s Oster was married, with two sons and a daughter.

2. The debate among scholars on German anti-Semitism and Nazi intentions toward the Jews continues. Recently Daniel Jonah Goldhagen's *Hitler's Willing Executioners: Ordinary Germans and the Holocaust* (New York, 1996) produced considerable debate. See the selection of critical articles in *Unwilling Germans? The Goldhagen Debate*, edited by Robert R. Shandley (Minneapolis, 1998).

3. Lucy S. Dawidowicz, *The War Against the Jews, 1933–1945* (New York, 1986), p. 191.

4. Schultze to Deutsch, August 19, 1972, in Deutsch Papers. Quotes in this paragraph are from this letter.

5. There is a complete copy of the Hossbach memorandum in German in the *Trial of the Major War Criminals before the International Military Tribunal, 14 November 1945–1 October 1946* (hereafter cited as IMT), Nuremberg (1946–49), xxv, pp. 403–13, Prosecution Exhibit 386 PS. See also *Documents on German Foreign Policy*, series D, vol. 2 (hereafter cited as *DGFP*), *DGFP-D-1*, no. 19, as quoted in Telford Taylor, *Munich: The Price of Peace* (New York, 1979), pp. 299–302. All quotes in this section are from Taylor.

6. Although Göring fell out of favor in 1942, he was Hitler's publicly designated successor from September 1, 1939, until the end of the Third Reich.

7. Blomberg was finally appointed as commander of the First Division in 1929, and served in that capacity until 1933.

8. On the strength of the Czech defenses, see Ivan Pfaff, "Die Modalitäten der Verteidigung der Tschechoslowakei 1938 ohne Verbündete," *Militärgeschichtliche Mitteilungen* 57 (1998), pp. 23–77.

9. All quotes from Taylor, *Munich*, pp. 298–307.

10. Andrew Roberts, *"The Holy Fox": A Biography of Lord Halifax* (London, 1991), pp. 9, 45. I am indebted to Roberts for the biographical details about Halifax. See also Lord Birkenhead, *Halifax: The Life of Lord Halifax* (London, 1961).

11. Halifax, "My Visit to Berlin, November 1937," May 6, 1946, Halifax Papers, York University. See also Halifax's document titled "This Memo. is

based on notes that I made in the train on my way back from Berlin," Halifax Papers, 410.3.3 (vi), and Sir Nevile Henderson's "Memorandum by His Majesty's Ambassador of a Conversation between Lord Halifax and Dr. Goebbels on November 21st, 1937," Halifax Papers 410.3.3 (vi). Halifax addressed this meeting, largely by quoting from his diary, in his memoirs, *The Fulness of Days* (London, 1957), pp. 183–91. Chatsworth was the hereditary estate of the dukes of Devonshire. Halifax's reference is meant to reflect Göring's interest in and knowledge of hunting of all kinds.

12. See *Die tödliche Utopie: Begleitband zur Dokumentationsstelle Obersalzberg*, edited by Horst Möller (Munich, 1999).

13. Taylor, *Munich*, p. 309.

14. Ivone Kirkpatrick, *The Inner Circle* (London, 1959), p. 97.

15. Ibid.

16. "My Visit to Berlin, November 1937," Halifax Papers.

17. Taylor, *Munich*, pp. 308–12. See also Roberts, *"The Holy Fox,"* pp. 64–75.

18. *DGFP*-D-I, no. 31, pp. 54–67, as quoted in Taylor, *Munich*, p. 312.

19. Paul Otto Schmidt, *Hitler's Interpreter* (New York, 1951), p. 76.

20. Taylor, *Munich*, p. 312.

21. Deutsch, *Hitler and His Generals.* pp. 82–96. All quotes are from this source.

22. Deutsch postwar interview with Wiedemann, quoted in *Hitler and His Generals*, p. 106. As Blomberg's situation became more widely known, according to Erich Schultze, wags began to refer to the "aging Fieldbed Marshal" (*alternden Feldbett-Marschall*). Schultze, "Die heimtückischen Verbrechen an Generaloberst Freiherr von Fritsch," n.d., in "Schultze Misc. Papers" file, Deutsch Papers.

23. Deutsch postwar interrogation of Blomberg, as quoted in *Hitler and His Generals*, p. 111.

24. Thun-Hohenstein, *Der Verschwörer*, p. 62. Gisevius had been in and out of postings in the Reich and Prussian Interior Ministries until the summer of 1936. By late 1937, he was no longer employed as a police official. See Hoffmann's introduction to *To the Bitter End* (1998 ed.).

25. Deutsch, *Hitler and His Generals*, pp. 161–66. All quotes are from these pages. See also Friedrich Hossbach, *Zwischen Wehrmacht und Hitler 1934–1938* (Göttingen, 1965), pp. 107–12.

26. Quoted in Heinz Höhne, *Canaris* (New York, 1979), p. 262.

27. Quoted in ibid., p. 258.

28. *Spiegelbild*, p. 430. The Fritsch crisis, according to Franz Maria Liedig, "strengthened Oster's [anti-Nazi] views and led him to take a deeper look at the whole problem."

29. Hjalmar Schacht, *Account Settled* (London, 1949), p. 113. Schacht served as minister of economics until November 26, 1937, but as minister "without portfolio," at Hitler's insistence, until January 22, 1943. He also served as president of the Reichsbank until January 1939.

30. Deutsch postwar interview of Schacht, as quoted in *Hitler and His Generals*, p. 242.

31. Ibid., p. 227. According to Deutsch, "The RM 80,000 figure was attested to Major Siewert by Captain von Both, Fritsch's adjutant, Siewert interview, January 15, 1970. In one of a string of perjuries committed by Brauchitsch at Nuremberg after the war, he denied that he had received any monetary gift from Hitler. IMT, XX, 583." Nevertheless, this gift and its transmission continue to be a subject of controversy. See Peter Hoffmann's review of Deutsch's work: "Harold C. Deutsch: Das Komplott oder Die Entmachtung der Generale. Blomberg- and Fritsch-Krise. Hitler's Weg zum Krieg," trans., Burkhardt Kiegeland (Zurich: Neue Diana-Press, 1974), in *Militärgeschichtliche Mitteilungen* (1976), vol. 2, pp. 196–201.

32. Statement of Curt Siewert, April 8, 1952, Institut für Zeitgeschichte [hereafter IfZ], ZS 148.

33. Gen. Curt Liebmann remembered what Hitler had said in his speech on February 4:

> His [the führer's] trust in the officer corps had suffered a heavy blow [from the Blomberg case]. In addition, Blomberg had never been the support he had hoped for. He very soon realized his weak character, and precisely for that reason did everything to strengthen his apparent authority. Göring can witness that the driving forward of rearmament was in no way Blomberg's deed, but his [Hitler's] own. In every critical situation, especially during the reoccupation of the Rhineland, Blomberg had lost his nerve.

Then Hitler turned to the Fritsch case, and spoke openly of the general's homosexuality. "Personal Experiences of General Curt Liebmann during 1938/39, recorded in November 1939," IfZ, ED 1. Simultaneously the press announced the creation of a new office, the Oberkommando der Wehrmacht (OKW), with General Keitel as chief, which would become Hitler's instrument of controlling the armed forces. The Army High Command, Oberkommando des Heeres (OKH), was relegated to a secondary role.

34. On the surface it appeared as though Hitler's decision to invade Austria was a preemptive strike sparked by Prime Minister Schuschnigg's call for a plebiscite on March 13 that would ask Austrians whether they endorsed "a free and German Austria." But its timing may have been

influenced by the führer's desire to avert attention from the Fritsch affair. According to Carl Goerdeler:

Neurath told me at the beginning of February one would now tackle the Austrian problem. Schuschnigg was expected in March. Now Hitler accelerated the affair to pacify the unrest in the Army by success. Schuschnigg had to come in February and was given an ultimatum. Papen [then the German ambassador in Vienna] perfidiously advised him to hold a vote. Hitler called this perfidious and invaded.

Bundesarchiv NL 113/000026 fol. 1, pp. 18–19. Captain Gerhard Engel, then newly arrived as Hitler's army adjutant, said, "I am still convinced that Hitler exaggerated the Austrian crisis because he needed a distraction." He cited Hitler's comment at that time, "I need finally another political success." Engel noted "That kind of talk occurred often. He [Hitler] was too smart to connect it [the Austrian invasion] with the Fritsch or Blomberg affairs, but purely intuitively one noticed that this internal connection was there." Engel interview with Harold Deutsch, May 11, 1970, pp. 2, 4, IfZ, ED 53.

35. Otto John, *Twice Through the Lines* (New York, 1972), p. 28.
36. Otto John spoke with Rundstedt in an Allied POW camp in 1947. The general defended his actions by saying, "On Hitler's instructions Himmler would never have accepted a duel and the affair would only have stirred up a lot of mud against the Army." Ibid., p. 29.
37. Public Record Office London, [hereafter PRO], Foreign Office [hereafter FO] 371/20733/251–63.
38. PRO, FO 371/20733/260. See also FO 371/20733/293–94.
39. Deutsch, "German Soldiers in the 1938 Munich Crisis," p. 313. Deutsch cites Oster, Canaris, and *Halder* as Goerdeler's sponsors, but, given Halder's distrust of the loquacious Goerdeler, it seems more likely that Beck was the sponsor from the General Staff.
40. The FO reports on Goerdeler's visits in spring 1938 are in the following files: PRO, FO 371/21662/39–41; FO 371/21662/82–84; and FO 371/21660/244–55. See also Ritter, *German Resistance*, pp. 84–85. In the Harold Nicolson manuscript diaries at Balliol Library at Oxford, there is the following entry for March 13: "Poor Altmayer—he has got here on a Jugo-Slav passport. He has come here on a mission which is so secret that I cannot even confide it to my diary. He is a hero." This entry was not included in *Harold Nicolson Diaries and Letters*, edited by his son Nigel, and published in 1966. It seems likely that Nicolson, through his excellent connections, was referring in code to Goerdeler. If word of his visit had spread beyond

Whitehall, it is no wonder that Goerdeler nervously told his British contacts that he came to them "with a noose around his neck." I am indebted to my student Janice Law for bringing this reference to my attention.

41. *DGFP*-D-II, no. 107, pp. 197–98, as quoted in Taylor, *Munich*, p. 380.

42. Nuremberg Document 588 PS, item 2, as quoted in Taylor, *Munich*, p. 388.

43. Roberts, *"The Holy Fox,"* p. 102.

44. Chamberlain to his sister, as quoted in Taylor, *Munich*, p. 655. Actually Chamberlain was misquoting Wellington on Waterloo. The duke described the battle as ". . . a damned nice thing—the nearest run thing you ever saw. . . ."

45. *The Diaries of Sir Alexander Cadogan, 1938–1945*, ed. David Dilks (Putnam: New York, 1972), p. 79, and Taylor, *Munich*, p. 655.

46. Schmidt, *Hitler's Interpreter*, p. 88.

47. Wiedemann affidavit, Nuremberg Document 3037 PS, as quoted in Taylor, *Munich*, p. 394.

48. Nicholas Reynolds, *Treason Was No Crime: Ludwig Beck, Chief of the German General Staff* (London, 1976), p. 151.

49. Ian Colvin, *Chief of Intelligence* (London, 1951), p. 55.

50. Schlabrendorff, *Secret War Against Hitler*, p. 91.

Chapter 2: The Road to Mutiny and Revolution

1. All quotes from Reynolds, *Treason Was No Crime*, pp. 118–20. There was a major scholarly controversy between Professors Müller and Hoffmann about Beck's true motives. Klaus-Jürgen Müller, *General Ludwig Beck: Studien und Dokumente zur politisch-militärischen Vorstellungswelt und Tätigkeit des Generalstabschefs des deutschen Heeres 1933–1938*, Schriften des Bundesarchivs no. 30 (Boppard, 1980); Peter Hoffmann, "Generaloberst Ludwig Becks militärpolitisches Denken," *Historische Zeitschrift* 234 (1982), pp. 101–21; Müller, "Militärpolitik, nicht Militäropposition! Eine Erwiderung," in *Historische Zeitschrift* 235 (1982), pp. 355–71. Müller's analysis of Beck's thought can be found in English in "Colonel-General Ludwig Beck, Chief of the General Staff, 1933–38," in his *The Army, Politics, and Society in Germany, 1933–45*, pp. 54–99.

2. Quoted in Höhne, *Canaris*, p. 264.

3. The entire memorandum is reprinted in Wolfgang Foerster, *Generaloberst Ludwig Beck: Sein Kampf gegen den Krieg* (Munich, 1953), pp. 100–105. Foerster was Beck's close friend and first biographer. See also Reynolds, *Treason Was No Crime*, pp. 149–50, and Hoffmann, *History of the German Resistance*, pp. 71–2. According to Beck, the only time he had presented his

views to Hitler face-to-face was for five minutes, when he had not been given time to prepare, at the time of the Anschluss. Foerster, *Beck*, p. 84.

4. Quoted in Reynolds, *Treason Was No Crime*, p. 153. See also Foerster, *Beck*, pp. 114–15.

5. Engel interview with Harold Deutsch, May 11, 1970, IfZ, ED 53, pp. 26–27.

6. For a full discussion of Vansittart's anti-Nazi (and anti-German) views, see Antoine Capet, "Deux regards antinomiques sur l'allemagne, 1933–1946," in *Contre le nazism ou contre l'Allemagne? Le débat sur l'anti-germanisme en Grande-Bretagne depuis la Deuxième Guerre mondiale*, edited by Antoine Capet and Jean-Paul Pichardie (Rouen, 1998), pp. 7–32.

7. Quoted in Ian Colvin, *The Chamberlain Cabinet* (London, 1971), pp. 132–33. On June 15 Halifax wrote to Dawson, mildly upbraiding him for the editorial recommending a plebiscite, and reminding him that HMG "would prefer a satisfactory settlement within the Czechoslovak State to an Anschluss with Germany." Halifax to Dawson, June 15, 1938. PRO, FO 800/309/183–84.

8. *DGFP-D-II*, no. 237, June 3, Henlein and Lorenz, as quoted in Taylor, *Munich*, p. 656.

9. Engel interview with Harold Deutsch, May 11, 1970, p. 8, IfZ, ED 53.

10. "Personal Experiences of Gen. Curt Liebmann during 1938/39," recorded in November 1939, pp. 2–3, IfZ, ED 1. All Liebmann quotes are from this source.

11. The morose general never recovered his equilibrium. He insisted on joining his regiment at the outbreak of war, and he was killed during the Polish campaign (September 1939) when he recklessly exposed himself to enemy fire. See W. A. Fletcher, " 'Dulce et Decorum est Pro Patria Mori:' The Dismissal and Death of Generaloberst Werner Freiherr von Fritsch," *University of Colorado Bulletin* 2 (1961), pp. 61–78, and Gerd Brausch, "Der Tod des Generalobersten Werner Freiherr v. Fritsch," *Militärgeschichtliche Mitteilungen* 1 (1970), pp. 95–112.

12. Edgar Röhricht, *Pflicht und Gewissen: Erinnerungen eines deutschen Generals 1932 bis 1944* (Stuttgart, 1965), p. 120f., quoted in Reynolds, *Treason Was No Crime*, pp. 154–155. See also Hoffmann, *History of the German Resistance*, p. 73.

13. Quoted in Foerster, *Beck*, p. 116.

14. H. C. T. Stronge, "The Czechoslovak Army and the Munich Crisis: A Personal Memorandum," in Brian Bond and Ian Roy, eds., *War and SOE* (New York, 1975), pp. 162–77. Copy of Stronge's article in the Deutsch Papers. All quotes are from this document.

15. Pfaff, "Die Modalitäten der Verteidigung der Tschechoslowakei 1938 ohne Verbündete," pp. 23–77.
16. Halifax apparently gave greater credence to the much more pessimistic reports of the British military attaché in Berlin, Colonel Frank Mason-Macfarlane. Stronge wrote, "It is unfortunate that his [Mason-Macfarlane's] views on Czech army morale would seem to have been accepted in Whitehall rather than mine." "The Czechoslovak Army and the Munich Crisis," p. 174.
17. NA T78, Roll 300, frame 6251277 ff., as quoted in Taylor, *Munich*, p. 687. Hitler originally intended for the western defenses to be manned by reservists (Landwehr),
18. General Adam's Memoirs, dictated January 24, 1940, IfZ, ED 109/5, p. 748.
19. Taylor, *Munich*, pp. 689–690. General Otto Förster recalled the führer's sudden change in priorities. Whereas in 1936 Hitler had estimated that the completion of the West Wall would take fifteen years, he revised his estimate to four years by 1937, "and in April 1938 it was: immediately!" "Landesbefestigung," April 12, 1957, IfZ, 1133.
20. Colonel General Wilhelm Adam, "Erinnerungen," vol. I, pp. 444–46, Deutsch Papers.
21. "Personal History of Erich Kordt, German Diplomat," Shanghai, September 15, 1945, in "Kordt mat." folder in Deutsch Papers. All quotes are from this document. Theo also mentions this meeting in "Timeline," attached to letter from Lore Kordt to Harold Deutsch, November 4, 1970, in Deutsch Papers. Conwell-Evans's relations with the FO were strained at this time, and may have contributed to his inability to get the message through. See T. Philip Conwell-Evans, *None So Blind: A Study of the Crisis Years, 1930–1939. Based on the private papers of Group Captain M[alcolm] G. Christie, Dr. Ing., formerly British Air attaché in Washington, Berlin, Stockholm, Oslo, Copenhagen.* (*Priv. printed*, London, 1947).
22. Quoted in Hoffmann, *History of the German Resistance*, p. 75. The entire memorandum is reprinted in Foerster, *Beck*, pp. 116–21. See also Reynolds, *Treason Was No Crime*, pp. 157–59 and Müller, *General Ludwig Beck*, pp. 551-554.
23. Halder in response to inteview by Josef Müller, Nov. 10, 1965, Bundesarchiv-Militärarchiv [Hereafter cited as BA-MA] N220/175. Halder to Krausnick, April 26, 1955, and Halder to Hermann von Witzleben, September 9, 1952, in "Halder Correspondence" folder in Deutsch Papers. Halder said that Oster complained to him about Beck's "lack of resolve," but Halder had the impression that Oster "did not

dare do anything decisive without Beck's agreement." Ibid. On Oster's influence on Beck, see Klaus-Jürgen Müller, *Das Heer und Hitler: Armee und nationalsozialistisches Regime 1933–1940* (Stuttgart, 1969), p. 325.

24. Fritz Wiedemann, *Der Mann der Feldherr werden wollte* (Velbert-Kettwig 1964), pp. 161–67. Cadogan's cryptic record of this meeting is in his *Diaries,* pp. 87–88.

25. Halifax's memoir of the conversation in PRO, FO 800/314/10–19. There is another copy in PRO, CAB [cabinet] 23/94/192–200. See Wiedemann's recollections about Ribbentrop's oafish behavior as ambassador. *Feldherr,* p. 168.

26. Ian Colvin, the British reporter with excellent sources among both Nazis and anti-Nazis, speculated that Hitler sent Wiedemann to London "to discover whether Halifax had altered his outlook and methods since the visit to Berchtesgaden [of November 1937]." *Chief of Intelligence,* p. 58.

27. Wiedemann, *Feldherr,* pp. 159–60.

28. PRO, CAB 23/94/163–64.

29. Wiedemann to Lord Halifax, March 6, 1947, Halifax Papers, A4.410.39.33.

30. Wiedemann, *Feldherr,* p. 171. Frau von Dohnanyi claimed that Wiedemann was the only member of Hitler's entourage whom the resisters trusted. She said that her late husband claimed that Hitler had said, in 1937, "Each generation needs its war and I will take care that this generation, too, will get its war." After that Wiedemann apparently expressed his dilemma: "I admit, here only a pistol will help, but who shall do it? I cannot help anybody murder someone who put his trust in me." Discussion with Frau Christine v. Dohnanyi, IfZ, ZS 603.

31. Reynolds, *Treason Was No Crime,* p. 159–60.

32. Ibid., p. 160, and Müller, *General Ludwig Beck,* pp. 554–56. All quotes in this section from these sources. Foerster, *Beck,* pp. 124–25, "Tcheka [or Cheka] methods" is a reference to the Russian precursor to the KGB. Beck meant to compare the Nazi reliance on the Gestapo and SD unfavorably to their Russian counterparts.

33. Quoted in Michael Bloch, *Ribbentrop* (London. 1992), p. 82.

34. Ernst von Weizsäcker, *Memoirs,* translated by John Andrews (Chicago, 1951), pp. 135–36. All quotes are from this source.

35. Chamberlain to his sister, as quoted in Taylor, *Munich,* p. 659.

36. Quoted in R.A.C. Parker, *Chamberlain and Appeasement: British Policy and the Coming of the Second World War* (New York, 1993), p. 151.

37. PRO, CAB 23/94/44.

38. PRO, CAB 23/94, as quoted in Taylor, *Munich,* p. 657.

39. *Documents on British Foreign Policy* [hereafter *DBFP*] 3dI, nos. 495, 508. 40.

40. Quoted in Parker, *Chamberlain and Appeasement*, p. 151. Nevile Henderson was an emotional and malleable man who was easily intimidated by Hitler. His tenure as ambassador (1937–39) was a disaster for British diplomacy. He consistently overreacted to Hitler's threats and was unable to present a balanced appraisal of changing situations to his government. Fabian von Schlabrendorff commented on Henderson's representation from the perspective of the resisters: *Secret War Against Hitler*, p. 93. For Sir Nevile Henderson's dubious self-evaluation of his work as ambassador, see his *Failure of a Mission: Berlin 1937–1939* (New York, 1940).

41. Foerster, *Beck*, p. 125.

42. Colvin, *Chief of Intelligence*, p. 61.

43. Foerster, *Beck*, pp. 125–28, as quoted in Hoffmann, *History of the German Resistance*, p. 77. See also Reynolds, *Treason Was No Crime*, p. 162.

44. Foerster, *Beck*, pp. 126–27, 136–37, as quoted in Hoffmann, *History of the German Resistance*, p. 77. Helldorf had also been a notorious SA member during the 1920s and 1930s. Ted Harrison, "'Alter Kämpfer' im Widerstand: Graf Helldorf, die NS-Bewegung und die Opposition gegen Hitler," *Vierteljahrshefte für Zeitgeschichte* 45 (1997), pp. 385–423.

45. Höhne, *Canaris*, p. 290.

46. Colvin, *Chief of Intelligence*, p. 60.

47. Adam, "Erinnerungen," vol. 1, pp. 452–53, Deutsch Papers. Adam specifies the date of this meeting only as a "Sunday in late July," which may have been July 24 but was likely July 31. See also Thun-Hohenstein, *Der Verschwörer*, p. 106.

48. Adam, "Erinnerungen," vol. 1, p. 454.

49. There was a difference of opinion about whether Beck or Brauchitsch read the memo. Professor Müller cites General Adam as recalling postwar that Beck read his own memo. Müller, *Beck*, p. 355, n. 138. Although General von Weichs recalled after the war that Brauchitsch read it, Müller thinks that highly unlikely because the timid Brauchitsch, although in agreement with Beck, would not so publicly have repudiated the führer's policies.

50. Adam, "Erinnerungen," vol. 1, pp. 455–66. Taylor, *Munich*, p. 693. See also Reynolds, *Treason Was No Crime*, pp. 162–64.

51. Major General von Weichs, quoted in Taylor, *Munich*, p. 693.

52. Adam, "Erinnerungen," vol. 1, p. 466. The general in question must have been Kluge; Liebmann claimed that he had not attended this meeting. "Personal Experiences of Gen. Curt Liebmann in the years 1938/39," IfZ, ED 1.

53. Erich Kordt to Harold Deutsch, February 14, 1947, "Kordt Corr (Simonis)" folder, Deutsch Papers. Kordt did not share Beck's optimism. "Hitler in my view could not be forced down by a 'generals' strike.'" Klaus-Jürgen Müller emphasizes the disunity and lack of consensus among the generals in his description of this meeting. *Das Heer und Hitler*, p. 336.

54. Reynolds, *Treason Was No Crime*, p. 164.

55. Kordt, *Nicht aus den Akten*, pp. 237–38.

56. Quoted in Taylor, *Munich*, p. 697.

57. Jodl's *Diary*, in IMT, xxviii, p. 374.

58. "Personal Experiences of Gen. Curt Liebmann in the years 1938/39," pp. 3–4, IfZ, ED 1, and Taylor, *Munich*, p. 698. All quotes are from these sources.

59. Quoted in Matthew Cooper, The *German Army 1933–1945: Its Political and Military Failure* (Lanham, Md., 1978), p. 98.

60. "Personal Experiences of Gen. Curt Liebmann in the years 1938/39," S. 419, IfZ, ED1.

61. Affidavit of Gen. Hans Doerr, January 3, 1952, IfZ, ZS 28.

62. Gisevius, *To the Bitter End*, pp. 280–81. In his introduction to Gisevius's book, Peter Hoffmann notes that Gisevius's account of events and characterization of fellow resisters is sometimes skewed, particularly of Colonel Claus von Stauffenberg (whom Gisevius intensely disliked) and his 1944 bomb plot against Hitler. Overall, however, Hoffmann writes that Gisevius's narrative is "more or less accurate," and his "account of the conspiracy to overthrow Hitler is important because it is, to a large extent, a rare firsthand account, detailed, and chronologically close to events, although it is colored by Gisevius's flamboyance and resentments," xx. Nearly everyone who worked with Gisevius considered him difficult. Erich Kordt found him "rather untractable [*sic*]. He is a little perhaps the type of a professional conspirator who finds it difficult to do anything normally." Erich Kordt to Harold Deutsch, Feb. 14, 1947, "Kordt Corr (Simonis)" folder, Deutsch Papers. Harold Deutsch, who got to know Gisevius just after the war, and corresponded with him until Gisevius's death, noted that he was difficult, but called him "an insightful and basically reliable witness." "German Soldiers in the 1938 Munich Crisis," p. 319.

63. Gisevius, *To the Bitter End*, pp. 300–304. All quotes are from this source.

64. Postwar correspondence between Dr. Helmuth Krausnick and General Halder throws some light on the origin of this trip. According to Krausnick, "Kleist's son recalls from his father's stories that the trip to England was his father's own idea, and that the trip, however, was undertaken with the full agreement of military intelligence (Abwehr) (Canaris and Oster) and their 'technical' assistance." Halder replied:

"Your statement that I had not been informed of the dispatching of Herr von Kleist-Schmenzin to London in August 1938 is correct. Neither Beck nor Oster informed me of it." Krausnick to Halder, April 20, 1955, and Halder to Krausnick, April 26, 1955, "Halder Correspondence" folder in Deutsch Papers.

65. Schlabrendorff, *Secret War Against Hitler*, p. 84.

66. Ian Colvin, *Vansittart in Office* (London, 1965), p. 223, and Colvin, *Chief of Intelligence*, p. 62.

67. Colvin described how difficult it was to get out of Berlin at this time: *Chief of Intelligence*, p. 63.

68. Schlabrendorff wrote of Colvin: "He was a member of the Casino Society, the most exclusive club in Berlin. Colvin had all the superior traits traditionally associated with the Englishman: He was intelligent, discreet, cautious, and at the same time daring. By arranging a visit to England by Ewald von Kleist, one of Hitler's most unbending adversaries, Colvin helped to establish a truly effective contact." *Secret War Against Hitler*, p. 91.

69. Colvin, *Chief of Intelligence*, p. 64–65.

70. Detlev Clemens, "The 'Bavarian Mussolini' and his 'Beerhall Putsch': British Images of Adolf Hitler, 1920–1924," *English Historical Review*, 114, no. 455 (February 1999), p. 73.

71. PRO, FO 371/21732/66.

72. PRO, FO 371/21732/70. Interestingly, Goerdeler had used a similar phrase about coming "with his neck in a halter" in his 1937 visit to England.

73. Vansittart had difficulty seeing the difference between Hitler and the conservative nationalist politicians who visited him. "Of all the Germans I saw," Vansittart confided in Colvin after the war, "Kleist had the stuff in him for a revolution against Hitler. But he wanted the Polish corridor, wanted to do a deal." *Chief of Intelligence*, p. 65. That is almost surely a misrepresentation of Kleist's statement. The relevant FO documents make no mention of Kleist wanting to do a deal, or even mentioning the Polish Corridor. But after the war Vansittart was very bitter toward all Germans, and refused to admit that he had ever had dealings with members of the German opposition. In his autobiography Vansittart merely mentioned Erich and Theo Kordt as Germans who, like Papen, "remained in Hitler's service." Lord Vansittart, *The Mist Procession* (London, 1958), p. 495.

74. Churchill's "Note of Conversation at Chartwell between Monsieur de K. and Mr. Winston Churchill, August 19. 1938," and Churchill's letter to Kleist are in the Churchill Archives, Churchill College, Cambridge, CHAR [Chartwell] 2/340B, 152–157. Churchill's cover note to Halifax,

and copies of his "Note," and his letter to Kleist are in PRO, FO 800/309/241–48.

75. Weizsäcker, *Memoirs*, pp. 137–38, 141–42. All quotes are from this source.
76. PRO, FO 800/314/59–60.
77. Quoted in Fest, *Plotting Hitler's Death*, p. 78. In a general FO memorandum, intended both for his colleagues and for Lord Halifax, Vansittart summarized information that he had garnered from his German informants. In particular he emphasized the views of "Herr X," whom Vansittart indentifies only as "a distinguished German economist," and who is undoubtedly Goerdeler. "Czechoslovakia: (Communications to Sir. R. Vansittart), 9th August 1938," PRO, FO 371/21736/ 181–211.
78. The British Foreign Office, as Detlev Clemens has recently shown, had been receiving reports about Hitler from its consul in Munich as early as 1920, and Hitler was regularly monitored from 1922 onward. Even in these early reports, Clemens stresses, British observers tried to separate Hitler from his "more radical lieutenants" and to convince themselves that it was possible to deal with him. As Clemens sees it: "Again and again Hitler's policy would prove this logic to be a misconception bordering on willful and fatal deception, the roots of which reached back into the early 1920s." 'Bavarian Mussolini.' p. 73.
79. PRO, FO 800/309/247.
80. PRO, FO 800/309/243–45.
81. "Memoir of Frau Christine v. Dohnanyi, née Bonhöffer," IfZ, ZS 603. Frau von Dohnanyi heard it from her husband, who in turn heard it from Minister of Justice Franz Gürtner, his immediate superior. See also Schlabrendorff, *Secret War Against Hitler,* p. 79.
82. Hitler quoted by Gerhard Engel in interview with Harold Deutsch, March 11, 1970, p. 25, IfZ, ED 53.
83. "Personal History of Erich Kordt, German Diplomat," 1945, in Deutsch Papers.
84. Gisevius, *To the Bitter End*, p. 282. Gisevius wrote:

> *Later on, I repeatedly asked Beck why he did not insist more energetically upon public announcement of his action. He did not deny that he had made a mistake, but he had never been one to thrust his own personality into the foreground. At that time he was still so deeply enmeshed in the traditions of the Prussian officers' corps that he wished to avoid even the faintest semblance of an attack upon the authority of the state. It was only with the passage of years that he outgrew these limitations.*

Ibid., p. 283.

85. Quoted in Reynolds, *Treason Was No Crime*, p. 170. On the relationship between Beck and Halder, and on Oster's attempt to enlist Halder to influence Beck, see H. G. Schall-Riaucour, *Aufstand und Gehorsam.— Offizierstum und Generalstab im Umbruch: Leben und Wirken von Generaloberst Franz Halder Generalstabchef 1938–1942* (Wiesbaden, 1972), pp. 214–15, 220. After the war Halder tended to exaggerate his role in the conspiracy while downplaying the roles of others. This earned him the enmity of his surviving contemporaries and the distrust of historians.

86. Hossbach, *Zwischen Wehrmacht und Hitler*, p. 136.

87. Liedig said of Beck's decision to resign and its effect on the Oster group: "The attempt by General BECK . . . to induce HITLER to abandon his plans to smash Czechoslovakia was bound to fail because BECK could not make up his mind to use force of arms. . . . It was, therefore, necessary for our group to intensify our activity with higher military personnel, and attempt to convince them that there was only one conclusive way to save the German people, namely, through the overthrow of the regime." Special Interrogation Report, 4 Oct. 45, p. 7, Deutsch Papers. See also Hermann Graml, "Der Fall Oster," *Vierteljahrshefte für Zeitgeschichte* (1966) p. 33.

88. The precise nature of Beck's activities after his resignation is unclear from the historical record. Ferdinand Sauerbruch, Beck's friend and fellow member of the prestigious Wednesday Society, related a conversation that he had with Beck in 1939. Beck claimed that he and Halder "had worked out plans for a coup d'état to coincide with an invasion of Czechoslovakia," and recounted them to the physician in great detail. *A Surgeon's Life* (London, 1953), pp. 238–39. Frau von Dohnanyi, Hans's widow, claimed that in the late fall of 1938, *after* the Munich conference, Dohnanyi had been forced out of the Ministry of Justice in Berlin by Martin Bormann, and exiled to Leipzig, Dohnanyi returned to Berlin regularly, in part to confer with Oster and Beck, who now became "the chief of the conspiracy." Every Thursday afternoon in the autumn of 1938, Dohnanyi and Oster drove to Beck's house in Lichterfelde for long conversations. Frau von Dohnanyi, "Dohnanyi, Oster, Mueller, Beck, Canaris," December 1, 1952, in Deutsch Papers. In August 1938 Hitler had offered Beck command of the First Army, but, in order to maintain the fiction that he had not retired as chief of the General Staff, Beck apparently never actually assumed command, although General Liebmann has Beck attending a meeting in Frankfurt-am-Main on September 2, conferring on West Wall defenses. "Personal Experiences of Gen. Curt Liebmann in the years 1938/39," p. 6, IfZ, ED 1. Beck wrote to Hossbach in October, telling him that Hitler had

reneged on the offer of the First Army command, claiming "that he expects that I bear the consequences of my actions in July and that he is of the opinion that the relationship built on faith as it must exist between him and the [commander] of the Army could not be restored in the required manner." Beck to Hossbach, 20 October 1938 BA-MA N24/29. According to Harold Deutsch, by the first week of September, Beck "was establishing treasonous foreign contacts in Basel." Cited in Peter Hoffmann, "Ludwig Beck: Loyalty and Resistance," *Central European History* 14, no. 4 (December 1981), p. 348. Beck retired from the army on October 19, 1938.

Chapter 3: Europe on the Edge of War

1. Gisevius, *To the Bitter End*, p. 324.
2. "Personal Experiences of Gen. Curt Liebmann for the years 1938/39," S. 419–420, IfZ, ED 1. All quotes from this source.
3. All Adam quotes from "Erinnerungen," vol. I, pp. 482–88. The contempt that Adam felt for Hitler did not go unnoticed by the führer. After Munich, Adam was unceremoniously dismissed from his post.
4. Jodl's *Diary* in IMT, xxviii, p. 375.
5. Ibid. Hitler's absorption in technical details and his autodidact's pride in this instance was expressed not only in his usual monologue, but in his memorandum on fortifications dated 1 July, 1938, *"Denkschrift zur Frage unserer Festungsanlagen,"* IMT, xxvii, 1801–1802 PS.
6. Adam was not the only person to comment on Hitler's table manners. Fabian von Schlabrendorff has left a vivid description of the führer at the table: *Secret War Against Hitler*, p. 234.
7. All quotes from Gisevius, *To the Bitter End*, pp. 283–86. Also Gisevius's testimony at the International Military Tribunal after the war: IMT, xii, 211–12. Gisevius is imprecise about the dates of the August and September meetings. Often the dates can be inferred with reasonable accuracy from internal evidence in the text. For example, Gisevius indicates that the three meetings with Halder all occurred when the general succeeded Beck as chief of the General Staff. Although Halder officially assumed the post on September 1, according to Beck's close friend and first biographer, Wolfgang Foerster, Halder actually took over Beck's office on Saturday, August 27. Foerster, *Beck*, p. 145. Given the urgent tone of this meeting, it is likely to have occurred on Halder's first day in office. Professor Peter Hoffmann is the first—and as far as I know, the only—historian to try to work out a chronology for the series of meetings that Gisevius described. For the most part I have followed Hoffmann's chronology. *History of the German Resistance,*

pp. 81–96, and especially supporting footnotes. When Hoffmann offers no guidance and the date is uncertain, I have made the best estimate.

8. Halder also had some reservations about Oster, whom he characterized as "an upstanding and respectable man, vivacious, imaginative, but terribly superficial." Halder also made reference to his "cheery thoughtlessness." The tension between the two, according to Oster's biographer, can be attributed to the difference between their two characters: "Halder the almost pedantic, hardworking general staff officer on the one hand and on the other Oster, elegant, in love with life, of a type more corresponding to a cavalry officer." Thun-Hohenstein, *Der Verschwörer*, pp. 33–34.

9. *DGFP*-D-II, No. 382, as quoted in Taylor, *Munich*, p. 664.

10. Colvin, *Chief of Intelligence*, pp. 66–67. It is unclear whether Kleist actually used the provocative term "preventive war" in conversation with the British. It does not appear in the British accounts of their meetings with him. If he had used the term, however, it may explain why Vansittart was so wary of him.

11. Helmuth Groscurth, *Tagebücher eines Abwehroffiziers 1938–1940* (Stuttgart, 1970), p. 104.

12. Theo Kordt, "Timeline," Deutsch Papers. Kordt noted only that the meeting took place "at the end of August 1938."

13. Klemperer, *German Resistance Against Hitler*, p. 23.

14. Kordt, *Nicht aus den Akten*, pp. 240–41 All quotes are from this source.

15. Ibid., pp. 241–244. All quotes are from this source.

16. "Personal History of Erich Kordt, German Diplomat," p. 8, Deutsch Papers. Erich Kordt was not the only conspirator who tried to sound out Brauchitsch. Elisabeth Gärtner-Strünck, who would later play a small but vital role in the conspiracy, was friendly with Gisevius. In the summer of 1938 the Strüncks were living at the Continental Hotel, where they became acquainted with their fellow resident, General von Brauchitsch. At an "accidental" meeting arranged by the Strüncks, Gisevius took Brauchitsch aside and asked to speak to him "about an important matter." Frau Strünck remembered that "Brauchitsch and Gisevius then went across to the clubroom and talked for three-quarters of an hour. We first waited in the hallway and then in our automobile parked in front of the hotel. Dr. Gisevius came out of the hotel. He told us the whole story and said that Brauchitsch was very open and Dr. Gisevius gained the impression that Brauchitsch had been won over for our matter." Frau Strünck to Helmuth Krausnick, 27 March 1963, IfZ, ZS 1811.

17. Kordt, *Nicht aus den Akten*, p. 244.

18. Technically it was a "meeting of ministers" rather than a cabinet meeting because of the lack of agenda and the absence of four ministers.

19. All quotes from PRO, CAB 23/94/285–318.

20. Field Marshal Manstein, testifying at Nuremberg, said "If it had come to war, we could not have defended either our west borders or the Polish border and, without a doubt, if Czechoslovakia had defended itself, we would have gotten stuck at its fortifications. We did not have the practical means to break through them." Quoted in Foerster, *Beck*, p. 84. William L. Shirer also analyzed the likely outcome of war in 1938: "Germany was in no position to go to war on October 1, 1938, against Czechoslovakia *and* France and Britain, not to mention Russia. Had she done so, she would have been quickly and easily defeated, and that would have been the end of Hitler and the Third Reich." *Rise and Fall of the Third Reich*, p. 520. Paul Johnson addressed this issue: "Would the Allies have been better advised to fight in autumn 1938 over Czechoslovakia, than in autumn 1939 over Poland? This too is in dispute; but the answer is surely 'Yes.'" *Modern Times: The World from the Twenties to the Nineties* (New York, 1991), p. 355. Most recently Graham Stewart has considered Chamberlain's role in the British assessment of war in September 1938: "Chamberlain was not blessed with hindsight [or], failing that, with superior wisdom to those who advised him. He could only draw up his plans on the basis of (flawed) information in his possession. This led him to believe that the balance of forces was more favourable to Germany in September than it would be later, an assessment that was, if anything, the reverse of the truth." *Burying Caesar*, p. 315.

21. See *Old Men Forget: The Autobiography of Duff Cooper* (New York, 1954), pp. 224–42, and John Charmley, *Duff Cooper: The Authorized Biography* (London, 1986), pp. 113–31. Other memoirs covering this period include L. S. Amery, *My Political Life: Volume Three—The Unforgiving Years 1929–1940* (London, 1959), pp. 259–95, R. J. Minney, *The Private Papers of Hore-Belisha* (London, 1960), pp. 135–54, and *The Memoirs of Anthony Eden, Earl of Avon: The Reckoning* (Boston, 1965), pp. 21–35.

22. Quoted in Taylor, *Munich*, p. 668.

23. Keith Feiling, *The Life of Neville Chamberlain* (London, 1947), p. 357, as quoted in Taylor, *Munich*, p. 669.

24. Churchill to Halifax, August 31, 1938. PRO, FO800/314/96–97.

25. Churchill continued to advocate a wider alliance in defense of the impending German attack on Czechoslovakia. On September 2 Ivan Maisky, the Soviet ambassador, drove to Chartwell to consult with Churchill about the Czech crisis. He offered to explore the possibility

of Britain, France, and the USSR coordinating their defense plans under the authority of Article 2 of the League of Nations Covenant, which obliged league members to consult if war threatened. The next day Churchill communicated Maisky's offer, but Halifax put him off. Martin Gilbert, *Churchill: A Life* (New York, 1991), p. 594.

26. Halder to Krausnick, April 26, 1955, in "Halder Correspondence" folder in Deutsch Papers:

> *Beck proposed sending Oberstleutnant (Lieutenant-Colonel) Böhm-Tettelbach, Ret., whom I also knew very well and who had very good personal relations to politically influential circles in London. The mission could only take place if I assumed responsibility for the mission and the selection of the person. I told Oster at the time that I was in agreement with the choice of the person and that I assumed responsibility for the mission in its entirety. I left the specifics of the instructions to Böhm-Tettelbach to Oster, because the specifics were dependent on the possibilities open to him in London.*

Halder also addressed this issue in "Aufzeichnungen zum Gespräch zwischen Herrn Generaloberst a. D. Halder und Dr. Uhlig am 2.6.53 in Königstein," in "Oppos-Brit." folder in Deutsch Papers. Böhm-Tettelbach claimed that he had first met with Oster and Halder on August 15 to plan the trip. Krausnick to Halder, 4 July 1955, in "Oppos-Brit." folder, Deutsch Papers. See the treatment of Böhm-Tettelbach's trip in Müller, *Das Heer und Hitler*, pp. 350–51.

27. Hans-Werner Böhm-Tettelbach, "Ein Mann hat gesprochen," *Rheinische Post*, July 19, 1948.

28. In 1946 Böhm-Tettelbach swore a deposition about his visit that was confirmed by a letter from Achim Oster, Hans's son, and a note from Gisevius. IfZ, ZS 633. There is some confusion about the dates in Böhm-Tettelbach's statement. He claims that he met with Oster on September 2 and also that he traveled to London by boat (*"nicht geflogen!"*) on September 2. It is possible that his meeting with Oster and the trip took place on the same day, but it seems more likely that he met Oster on September 1 and traveled on September 2.

29. Carl J. Burckhardt, *Meine Danziger Mission 1937–1939* (Munich, 1960), pp. 181–83. Weizsäcker also mentions this meeting, although he mistakenly places it "at the end of August." *Memoirs*, pp. 146–47.

30. Weizsäcker, *Memoirs*, p. 143.

31. "Interview with Dr. Schultze," 5 Dec., 1969, pp. 23, 30. Deutsch Papers. Schultze remembers this meeting taking place in the early fall.

32. "Besprechung am 3.9.38 auf dem Berghof," IMT, xxv, pp. 462–64.

33. Kordt, *Nicht aus den Akten*, p. 248–49. Kordt does not date this meeting, but from internal evidence it had to have taken place between September 2 and 4.

34. Ibid. Kordt obtained Weizsäcker's permission for the mission. The state secretary recounted the situation in his *Memoirs*, p. 145.

35. Gisevius, *To the Bitter End*, p. 287, and Gisevius's testimony at IMT. In the latter Gisevius claimed that the meeting took place "at the end of July 1938." This date is surely a mistake given the assurance with which Halder spoke of being in charge of bringing about "a revolt." IMT, xii, p. 212.

36. There remains some question about the extent of Beck's leadership role after his retirement. In 1952, at a discussion among Frau von Dohnanyi, General Hermann von Witzleben, General Edwin Lahousen, and General Kurt Sendtner, this issue was addressed. The meeting seems to have been prompted by the participating individuals' annoyance with Halder's derogatory comments about Beck—"a pure fool." General Witzleben said, "Beck was still the one, even after his retirement." Frau Dohnanyi agreed with him: "Nothing was done until Beck was informed and nothing would have happened if Beck had disagreed." "Discussion with Frau von Dohnanyi on 1 December 1952," IfZ, ZS 603. These recollections seem to me to be the result of nostalgic regard for Beck and annoyance with the difficult Halder. On Halder's denigration of Beck, see Halder Papers, BA-MA N220/175.

37. Gisevius, *To the Bitter End*, pp. 288–296, and Gisevius's testimony, IMT, xii, pp. 212–13. All quotes are from these sources.

38. Theo Kordt to Lord Halifax, December 17, 1947, Halifax Papers, A4.410.42 3. Kordt was writing to Halifax to see if the latter could intervene on behalf of Weizsäcker, who was being tried as a war criminal. Theo Kordt also repeated this statement, in a slightly shorter version, in "Persönliche Aufzeichnungen von Dr. Theo Kordt, Manuscript No. 1," undated, p. 6, in Deutsch Papers. Kordt, *Nicht aus den Akten*, pp. 279–81. All quotes are from these sources. Cadogan mentioned it in his diary, noting that "H saw 'Herr X' [Theo Kordt], today, and latter repeated his story [that is, the one he had given to Sir Horace Wilson the day before]. Wants us to broadcast to German nation. I said that is fatal—and the suggestion almost makes me suspect Herr X." *Cadogan Diaries*, p. 95.

39. After the Munich conference, Halifax admitted that he had misled Theo Kordt about the possibility of a strong British statement at their meeting of September 7: In a postwar letter, Kordt recounted Halifax's words to him: "We were not able to be so frank with you as you were with us. At the time in question we were already considering a per-

sonal initiative from the Prime Minister to Hitler explaining to him in an unqualified manner the British point of view." Theo Kordt to Lord Halifax, December 17, 1947, Halifax Papers, A4.410.42.3, p. 4.

40. Quoted in Taylor, *Munich*, p. 671. All quotes are from this source.

41. Gisevius, *To the Bitter End*, p. 304.

42. Gisevius claimed, "I kept pressing Oster until he established contact with [Witzleben]." *To the Bitter End*, p. 304. Gisevius periodically tended to exaggerate his role, and that is probably the case here. Almost surely Oster did not need Gisevius's prodding to make contact with a man he already knew. Schacht claimed that *he* had recruited Witzleben for the conspiracy in May 1938. *Account Settled,* p. 119. See also Peterson, *Hjalmar Schacht,* p. 315. Gisevius supported this view in his IMT testimony as a witness for Schacht: "Schacht won Witzleben over." xii, p. 214. It is likely that Oster knew about Witzleben's commitment to the resistance, and was just waiting to call on him when the plans became fixed.

43. Gisevius, *To the Bitter End*, p. 304.

44. Jodl's diary, IMT, xxviii, p. 376.

45. "Personal Experiences of Gen. Curt Liebmann for the years 1938/39," S. 421–22, IfZ, ED 1. Liebmann says only that this meeting took place during his stay in Nuremberg, September 5–11.

46. Ibid. In the first version of his notes, made in 1939, Liebmann omitted Halder's reply, writing, "I am afraid to put his response on paper." He added Halder's reply in a footnote to the original in 1947. It is uncertain whether Fritsch was aware of the conspiracy, and if so, whether he had pledged to support it. Certainly he did not have a leadership role in it.

47. *The Memoirs of Anthony Eden, Earl of Avon: The Reckoning* (Boston, 1965), p. 25. David Carlton has claimed that Eden's criticism of the prime minister was relatively mild, saying that Eden's position "was throughout marginally more resolute than that of Chamberlain." *Anthony Eden: A Biography* (London, 1981), p. 137. That seems to me to misrepresent the significant and growing gap between them through the month of September.

48. Henderson to Sir Horace Wilson, September 10, 1938, PRO, FO 800/314/134–35. On the earlier messages that Henderson had delivered—on August 31, September 1, and September 9—see Peter Hoffmann, "The Question of Western Allied Co-operation with the German Anti-Nazi Conspiracy, 1938–1944," *Historical Journal* 34 (1991), p. 439.

49. See Henderson's version of the incident in *Failure of a Mission*, p. 150, and Oliver Harvey's recounting of it in *The Diplomatic Diaries of Oliver Harvey 1937–1940* (London, 1970), pp. 174–75. When Eden discovered that Halifax had actually drawn up the recommended warning but not

sent it, he assumed that it was because of Henderson's advice. He wrote in his diary for September 10, "If this advice was given I believe it to be wrong." Eden, *The Reckoning*, p. 26. On September 11 Eden saw Halifax for the second time in three days and renewed his recommendation for a strong stance against German aggression, including overt acts of naval preparedness, which would get the attention of the Nazis more effectively than verbal warnings. The next day *The Times* published a letter from Eden that reiterated many of the themes that he had developed in his private communications with Halifax. And he unknowingly echoed a point made by Theo Kordt in the latter's discussion with Halifax five days earlier, when he wrote: "We have often been told that the war of 1914 would never have come about had the attitude of this country been clearly understood in time. Whatever we think of this statement it is the duty of each one of us, press and public as well as Government, to take every step in our power to prevent such a repetition of such tragedy." Ibid., p. 27.

50. "Besprechung Nürnberg 9./10.9.38 22,00—3,30 Uhr," IMT, xxv, pp. 464–69.

51. Gisevius, *To the Bitter End*, p. 319. All quotes are from this source.

52. Kordt, *Nicht aus den Akten*, pp. 254–255. Conwell-Evans did come back to London bearing "passionate pleas from moderate German leaders begging H.M.G. to take some step to stop their mad Chancellor!" *Diplomatic Diaries of Oliver Harvey*, p. 179.

53. Gisevius, *To the Bitter End*, pp. 305–17, and IMT, xii, p. 214.

54. The spearhead of the Twenty-third was Infantry Regiment nine, a unit whose officer corps was so dominated by conservative aristocrats that it was referred to as "I.R. von 9."

55. Gisevius, *To the Bitter End*, p. 306.

56. Ibid.

57. Erich Kordt to H. Deutsch, Feb. 14, 1947, in "Kordt Corr (Simonis)" folder, Deutsch Papers.

58. Gisevius, *To the Bitter End*, p. 306.

59. Ibid., pp. 306–07.

60. Ibid., p. 310.

61. Ibid.

62. Ibid., p. 311.

63. Ibid.

64. Ibid.

65. Ibid., p. 314.

66. Ibid., p. 315.

67. Ibid.

68. Hoffmann, *History of the German Resistance*, p. 89.

69. Transcript of a taped conversation between Frau Ursula von Witzleben and Harold Deutsch, February 10, 1970, Deutsch Papers. All quotes are from this document.

70. Gisevius does not give a date, but places this event just before Hitler's speech to the Nazi Party rally, which took place on September 12. *To the Bitter End*, p. 320, and IMT, xii, pp. 214–15.

71. Gisevius, *To the Bitter End*, p. 320; Frau Strünck, "Aktennotiz," 20 April 1964, IfZ, ZS 1811. All quotes are from these sources. Neither source specifies precisely which buildings in "Wilhelmstrasse" were targets. The list certainly included the Reich Chancellery, and almost surely the SS Headquarters on Prinz Albrecht Strasse.

72. Hossbach, *Zwischen Wehrmacht und Hitler*, p. 136. Hossbach claimed that Witzleben made the remark to him when he went to say good-bye to the general "due to my transfer to Göttingen in September 1938."

73. Kordt to H. Deutsch, Feb. 14, 1947, "Kordt Corr (Simonis)" folder, Deutsch Papers.

74. All quotes are from Norman H. Baynes, *The Speeches of Adolf Hitler: April 1922–August 1939* (New York, 1969), vol. 2, pp. 1487–1499.

75. Jodl's diary, IMT, xxviii, pp. 377–78.

76. Schmidt, *Hitler's Interpreter*, p. 90.

77. Franz Halder to Dr. Krausnick, April 26, 1955, in "Halder Corr." folder, Deutsch Papers.

78. Gisevius, *To the Bitter End*, pp. 298–300. Following quotes are from this source. Both in his book and in his postwar testimony, Gisevius claimed that there was an interval of "several weeks" between his first and second interviews with Halder. However, it is clear from the internal evidence that the second meeting had to have taken place before September 14. IMT, xii, p. 213.

79. IMT, vol. 12, p. 213.

80. Gisevius, *To The the Bitter End*, p. 300.

81. Kordt, *Nicht aus den Akten*, p. 258; Weizsäcker, *Memoirs,* pp. 142–143.

82. Gisevius, *To the Bitter End*, p. 321.

83. Theo Kordt, "Timeline," Deutsch Papers.

84. Groscurth, *Tagebücher*, p. 35.

85. All quotes from PRO, CAB 23/95/32–61.

86. Quoted in Taylor, *Munich*, p. 730.

87. Schacht, quoting Halder's postwar testimony, in *Account Settled*, p. 122.

88. Colvin, *Chief of Intelligence*, p. 69.

89. Kordt, *Nicht aus den Akten*, pp. 258–59.

90. Ibid., pp. 281–82.
91. Quoted in William Manchester, *The Last Lion: Winston Spencer Churchill: Alone: 1932–1940* (New York, 1988), p. 334.
92. Ibid. See also Gilbert, *Churchill*, p. 595.
93. Gisevius, *To the Bitter End*, pp. 321–22. Quotes from this source.

Chapter 4: Hitler's Knife at Chamberlain's Throat

1. *DGFP*-D-II, no. 486, and *Daily Herald*, Sept. 15, 1938, as quoted in Taylor, *Munich*, p. 732.
2. Quoted in ibid., p. 746–47.
3. All quotes from PRO, CAB 23/95/63–111.
4. Adolf Hitler interview with G. Ward Price in Baynes, *Speeches of Adolf Hitler*, p. 1500.
5. Ibid., p. 1503.
6. "Special Interrogation Report," 4 Oct. 1945, p. 2, in Deutsch Papers.
7. Heinz is the main source for the political goals of the conspirators. "Von Canaris zur NKWD," [1949] copies in IfZ and U.S. National Archives, p. 101; and Ritter, *The German Resistance*, p. 104, in quoting private correspondence from Heinz on this issue. Susanne Meinl claims that there were three distinct but overlapping political groups among the conspirators. Susanne Meinl, *Nationalsozialisten gegen Hitler: Die nationalrevolutionare Opposition um Friedrich Wilhelm Heinz* (Berlin, 2000), p. 294.
8. The main sources for Heinz's life and career are Meinl, *Nationalsozialisten gegen Hitler*, and Heinz's typescript autobiography, "Von Canaris zur NKWD."
9. The raiding party included student leaders Junker and Hoffmann, Albrecht Erich Günther from the magazine *Deutsches Volkstum*, Konrad Count Finckenstein, Hans-Jürgen Count Blumenthal, Haubold Count Einsiedel, Baron Treusch von Buttlar-Brandenfels, Lieutenant Hans-Albrecht Herzner (who, a year later, would lead the commando attack on the Jablunka Pass in Poland that opened World War II), Lieutenant Arnold Bistrick, from the Stahlhelm, Lieutenant Wolfgang Knaack and Lieutenant Franz Maria Liedig, both from the Abwehr, and Heinz. Thun-Hohenstein, *Der Verschwörer*, p. 108.
10. All quotes from Rainer Hildebrand, *Wir sind die Letzten: Aus dem Leben des Widerstandkämpfers Albrecht Haushofer und seiner Freunde* (Berlin, 1951), p. 93.
11. Schall-Riaucour, *Aufstand und Gehorsam*, p. 203.
12. Peter Hoffmann, *Hitler's Personal Security* (London, 1979), pp. 157–160.
13. Thun-Hohenstein, *Der Verschwörer*, pp. 109–11. Thun-Hohenstein, citing Hoffmann, claims that this "last" meeting of the Oster circle took

place sometime between September 15 and September 20. Given the uncertainty surrounding the Berchtesgaden meeting of September 15, and the relative confidence of the conspirators at this meeting, it is likely that it took place on or just before the twentieth. See also Hoffmann, *History of the German Resistance*, pp. 31–32, 92–93; Heinz, "Von Wilhelm Canaris zur NKWD," pp. 98–104. The Abwehr could draw on serving Wehrmacht officers for temporary duty, claiming "special operations."

14. All quotes from PRO, CAB 23/95/140–65.

15. Ibid. See also the *Diplomatic Diaries of Oliver Harvey*, p. 191.

16. *DGFP*-D-II, no. 570, as quoted in Taylor, *Munich*, p. 805.

17. Kordt, *Nicht aus den Akten*, p. 262.

18. Ibid., pp. 262–63. All quotes are from this source.

19. Gisevius, *To the Bitter End*, 323–24.

20. Ibid., p. 323.

21. Ibid.

22. "Personal History of Erich Kordt, German Diplomat," p. 8, Deutsch papers.

23. Gisevius, *To the Bitter End*, p. 324.

24. PRO, PREM [Premier: Prime Minister] 1/266A, No. 22, as quoted in Taylor, *Munich*, pp. 813–14.

25. All quotes from PRO, CAB 23/95/166–92. See also the dramatic account of the Bad Godesberg meeting by Paul Schmidt, in *Hitler's Interpreter*, pp. 95–102.

26. British military intelligence, under the spell of Colonel Charles Lindbergh, American air pioneer and purported expert on relative air force strengths in Europe, believed that in the event of war, Prague, Paris, and London would all be subjected to massive bombing by the Luftwaffe. In fact German bombers were designed as support for ground attacks and were mostly dive-bombers and medium range level bombers. Taylor says of the prospect of German bombers flying the six-hundred-mile round-trip mission (or eight-hundred-mile-trip if the Low Countries' neutrality were respected) to London and back: "These distances were at or beyond the extreme loaded range of the German bombers in 1938." *Munich*, p. 850. (The Heinkel 111, the workhorse medium bomber of the Luftwaffe in 1938, had a maximum range of 745 miles.) It is unlikely that the inaccurate estimates by British military intelligence affected Chamberlain's strategy in September 1938. He was already committed to peace at any price. But they did have currency among politicians and were circulated among decision makers, and thus contributed to the general air of defeatism.

27. *Cadogan Diaries*, p. 105.

28. Ibid., p. 103.

29. Ibid. Cadogan was not the only person trying to influence Halifax. Just prior to the cabinet meeting, Anthony Eden had left him a message urging that Halifax reject Hitler's Bad Godesberg ultimatum. Eden, *The Reckoning*, p. 32. Eden's biographer speculated that Eden's communication "may have helped to push Halifax into diverging from the Prime Minister, who favoured in effect surrendering to Hitler." Carlton, *Anthony Eden*, p. 139. See also David Dutton, *Anthony Eden: A Life and Reputation* (London, 1997), pp. 126–27. Leo Amery also left a note for Halifax on September 24, "deploring the idea of any pressure being put upon the Czechs to accept Hitler's ultimatum of the brutality and cunning of which I had already had some idea." *The Empire at Bay: The Leo Amery Diaries 1929–1945*, edited by John Barnes and David Nicholson (London, 1988), p. 514.

30. PRO, CAB 23/95/198–99.

31. Ibid. 200. Following quotes are from this source.

32. Ibid.

33. The notes are in the Halifax Papers, A4.410.3.7. All quotes are from this source. In his memoirs Halifax did not address his role during September 1938 in specific detail. In retrospect he thought that Chamberlain had explored all the alternatives to war, and had brought a unified country and the Commonwealth into war. "And that was the big thing that Chamberlain did." *Fulness of Days*, p. 198.

34. During an interval between cabinet meetings, Oliver Harvey encountered Hore-Belisha, who "was very stiff and bellicose. He felt the proposals must be rejected and now was the time to fight Hitler." *Diplomatic Diaries of Oliver Harvey*, p. 197.

35. All quotes from *DGFP*-D-II, no. 1092, as quoted in Taylor, *Munich*, pp. 829–30.

36. All quotes from PRO, CAB 23/95/234–45. Technically Daladier was "minister president."

37. Höhne, *Canaris*, p. 303.

38. Franz Maria Liedig, an Abwehr officer since 1936 and a close collaborator of Oster's, claimed that Canaris consciously chose to remain in the background. "His estimate of his own personality, his capabilities and his limitations restrained him from usurping such a [leadership] position for himself." "Special Interrogation Report, no. 6," 4 October 1945, Deutsch Papers.

39. Winston S. Churchill, *The Second World War*, vol. 1, *The Gathering Storm* (New York, 1961) p. 276–77.

40. Ibid.

41. Foreign Office release, *Documents on International Affairs*, vol. 2, p. 241, as quoted in Taylor, *Munich*, p. 863.

42. Private letter from Halifax to Churchill, July 24, 1947, as quoted in Taylor, *Munich*, p. 863.

43. Kirkpatrick, *Inner Circle*, p. 123, and Schmidt, *Hitler's Interpreter*, p. 103.

44. Schmidt, *Hitler's Interpreter*, p. 103.

45. Kirkpatrick, *Inner Circle*, p. 123.

46. All quotes from William L. Shirer, *The Nightmare Years, 1930–1940* (New York, 1984), pp. 349–51.

47. Colvin, *Chief of Intelligence*, p. 72.

48. Schmidt, *Hitler's Interpreter*, pp. 104–05.

49. Kirkpatrick, *Inner Circle*, p. 126.

50. Kordt, *Nicht aus den Akten*, p. 265.

51. All quotes from Röhricht, *Pflicht und Gewissen*, quoted in Taylor, *Munich*, p. 868.

52. Ruth Andreas-Friedrich, *Berlin Underground, 1938–1945* (New York, 1989), p. 2.

53. Shirer, *The Nightmare Years*, pp. 352–53.

54. Gisevius, *To the Bitter End*, p. 324. According to Gisevius, Witzleben said that he "would like to have had the guns unlimbered in front of the Reich Chancellery." Gisevius's testimony in IMT, xii, p. 219.

55. Schmidt, *Hitler's Interpreter*, p. 105.

56. Ibid. Erich Kordt noted this same exchange. He remembered it as "a dreary day," and noted that the Berliners "stood by in icy silence." "Sudeten Crisis" [from Erich Kordt], undated, in Deutsch Papers. Wiedemann also commented on Goebbels's role, noting that the minister of propaganda "was the most cynical but definitely the smartest of all of them." *Feldherr*, p. 177.

57. All quotes from *Cadogan Diaries*, pp. 106–9.

58. All quotes from PRO, CAB 23/95/260–78.

59. Colvin, *Chief of Intelligence*, p. 72. Oliver Harvey remembered September 27 as being a day when many Englishmen, himself included, steeled themselves for war. In a conversation with a colleague at the Foreign Office, he had said:

> *War is being forced on us by Hitler who seems determined not to have a peaceful settlement . . . Yet war may mean great surprises for us. Effect of bombing is unknown factor. German internal regime may crash. Czechoslovakia may resist longer than we expect. I said I felt something of the crusading spirit: we cannot have peace while regimes such as Hitler's and Mussolini's exist.*

Diplomatic Diaries of Oliver Harvey, p. 200.

60. Heinz, "Von Wilhelm Canaris zur NKWD," p. 98.

61. Erwin Lahousen, undated typescript (1945?), p. 2, in U.S. National Archives, as quoted in Hoffmann, *History of the German Resistance*, p. 93.

62. Heinz, "Von Wilhelm Canaris zur NKWD," pp. 99–100.

63. Thun-Hohenstein, *Der Verschwörer*, p. 115.

64. Gisevius, *To the Bitter End*, pp. 325–26, and IMT, xii, p. 219. According to Kordt, when Brauchitsch heard about the apparent English resolve to go to war, he shouted "He [Hitler] has lied to me again." *Nicht aus den Akten*, p. 269.

65. Schlabrendorff, *Secret War Against Hitler*, p. 102.

66. "Personal History of Erich Kordt, German Diplomat," 1945, p. 8, in Deutsch Papers.

67. All quotes from Kordt, *Nicht aus den Akten*, p. 270.

68. Jodl's *Diary* in IMT, xxviii, p. 388.

69. Wiedemann, *Feldherr*, p. 177.

70. Schmidt, *Hitler's Interpreter*, pp. 105–6.

71. Ibid., p. 106

72. Ibid., p. 107; Kordt, *Nicht aus den Akten*, p. 271.

73. Jodl's *Diary* in IMT, xxviii, p. 389.

74. All quotes from Taylor, *Munich*, pp. 10–11.

75. Manchester, *The Last Lion*, p. 347.

76. Gisevius, *To the Bitter End*, p. 325; Kordt, *Nicht aus den Akten*, p. 278.

77. Witzleben, quoted in Heinz, "Von Canaris zur NKWD," p. 101.

78. Chamberlain to his sister, as quoted in Taylor, *Munich*, pp. 64–65. The first time to which Chamberlain referred was Benjamin Disraeli's return in 1878, after having signed the Treaty of Berlin, which concluded the Russo-Turkish War. "Dizzy" had brought back what he termed "peace with honour."

79. PRO, CAB 23/95/279–89.

80. *Diplomatic Diaries of Oliver Harvey*, p. 202.

81. Quoted in Shirer, *Rise and Fall of the Third Reich*, p. 522.

82. Kordt, *Nicht aus den Akten*, p. 278.

83. IMT, xii, p. 220.

84. Gisevius, *To the Bitter End*, p. 326.

85. All quotes from Manchester, The *Last Lion*, pp. 364–69.

86. "Personal History of Erich Kordt," p. 8; Gisevius, *To the Bitter End*, p. 326. Even Sir Nevile Henderson, the arch-appeaser, agreed with this judgment. On October 6, 1938, he wrote to Halifax, saying, "by

keeping the peace, we have saved Hitler and his regime." *DGFP*-D-II, p. 615, quoted in Hoffmann, *History of the German Resistance*, p. 96.

Epilogue

1. "Interview with Attorney Liedig," August 1960, Deutsch Papers.
2. Testimony of Sas after the war before a Dutch board of inquiry, IfZ, 1626, p. 20, quoted in Klemperer, *German Resistance Against Hitler*, p. 195.
3. See Professor Deutsch's interview with Sas, undated, in "Sas & Oster," in "Oster-Sas-Goethals" file, Deutsch Papers.
4. IfZ ZS 1626, p. 9, quoted in Klemperer, *German Resistance Against Hitler*, p. 196.
5. Gisevius, *To the Bitter End*, p. 375.
6. Müller to Heinz Schmalschläger, 21 April 1970, "B-F Corr." file, Deutsch Papers. The following quotes in this paragraph are also from this source. See also Müller's letter to Cmdr. O. Benninghof (ret.), May 12, 1970, in the same file, which also discusses his mission to the Vatican.
7. On the entire Müller mission, see his report in National Archives microfilm T84, Roll 229, item EAP 21-x-15/2., as quoted in Hoffmann, *History of the German Resistance*, pp. 158–64. See also Müller, *Bis zur letzten Konsequenz*, and Josef Müller papers, IfZ, ED 92. Müller survived the war to become one of the cofounders of the Bavarian CSU party.
8. Roberts, "*The Holy Fox*," p. 184. See also Hoffmann, *History of the German Resistance*, pp. 155–58, and Ulrich von Hassell, *Die Hassell-Tagebücher 1938–1944: Nach der Handschrift revidierte und erweiterte Ausgabe*, rev. ed., edited by Friedrich. Hiller von Gaertringen (Berlin, 1988). Published in an earlier edition in English as *The von Hassell Diaries 1938–1944* (London, 1948).
9. It is clear from Theo Kordt's papers that Conwell-Evans came to Switzerland on October 24, and they met for the first time the next day. "Persönliche Aufzeichnungen von Dr. Theodor Kordt," undated, and Conwell-Evans to T. Kordt, October 20, 1939, in Deutsch Papers. In addition, they met three times between December 18 and February 16. Hoffmann, *History of the German Resistance*, pp. 154–55.
10. "Persönliche Aufzeichnungen von Dr. Theodor Kordt," Deutsch Papers.
11. Müller's evidence at Huppenkothen trial, quoted in Hoffmann, *History of the German Resistance*, p. 162. On British demands, see Hoffmann, "The Question of Western Allied Co-operation with the German Anti-Nazi Conspiracy," p. 451.
12. "Personal History of Erich Kordt, German Diplomat," p. 11, Deutsch Papers.

13. Kordt, *Nicht aus den Akten*, p. 371. Erwin Lahousen remembered that Oster had called him into his office one day in November: "I was asked by Colonel Oster to provide a special explosive . . . [since] an assassination of Hitler was intended." Lahousen confirmed that assassination attempt was to take place just before the launch of the western offensive, then planned for November 12. "A man who had access to Hitler was ready to throw the bomb. I later learned from various sources that it was the then legal advisor from the Foreign Office, Dr. Erich Kordt." Lahousen affidavit, 1 July 1947, IfZ, 658.

14. Kordt, *Nicht aus den Akten*, p. 371, as quoted in Hoffmann, *History of the German Resistance*, p. 256.

15. Peter Hoffmann has characterized the dynamic of the situation in the winter of 1939–40:

    The opposition group centred on Beck and Oster continued to lose ground both with Brauchitsch and with Halder. Both were too weak to follow matters to their logical conclusion and adopt a definitive attitude. They could not make up their minds either to fulfil their military duty in the true sense of the word or to do their ethical and moral duty. They really wanted to do both but then again they wanted neither. It was all too uncertain and too risky.

    History of the German Resistance, p. 146.

16. *Spiegelbild*, p. 370. Pfuhlstein claimed that while the death of Oster's son made him "despondent," he "continued to criticize [Hitler's] military measures openly."

17. Schlabrendorff, *Secret War Against Hitler*, pp. 229–39.

18. Dawidowicz, *War Against the Jews*, p. 191.

19. Gisevius, *To the Bitter End*, pp. 473–477; Schlabrendorff, *Secret War Against Hitler*, pp. 241–43. See also Christine v. Dohnanyi, "Aufzeichnung über das Schicksal der Dokumentensammlung meines Mannes, des Reichsgerichtsrats a. D. Dr. Hans von Dohnanyi," undated [1955?], IfZ, 603. Copy in Deutsch Papers. See also Winfried Meyer, *Unternehmen Sieben* (Frankfurt am main, 1993), and Eberhard Bethge, *Dietrich Bonhoeffer* (New York, 1977).

20. The Gestapo was only one of the several offices of the Reich Central Security Office (RSHA) whose *"Chef"* position Kaltenbrunner inherited from Heydrich only after several months' interim.

21. *Spiegelbild*, p. 287.

22. Höhne, *Canaris*, pp. 487–54.

23. "Woman as Consul," *Vancouver Province*, September 16, 1965. "New German Consul First Woman in Post," *Vancouver Sun*, September 25, 1965.

24. There is some interesting postwar correspondence among Theo Kordt, Halifax, and President Harry Truman, in which Kordt asks Halifax to intervene on behalf of Weizsäcker, which he did. See in particular the letter from Truman to Halifax, February 11, 1950. When Weizsäcker was released, he sent a note of thanks to Halifax. Weizsäcker to Halifax, December 6, 1950, Halifax Papers.

25. Heinz provides a dramatic description of his last meeting with Oster and Canaris, and an account of how they helped him escape:

> *After July 20th, . . . [in t]he armed forces central prison at Berlin Lehrterstrasse and the Gestapo bloody cellar in Prinz-Albrecht-Strasse I saw . . . Wilhelm Canaris and Hans Oster for the last time. [We] . . . were allowed to talk to one another . . . for a few minutes. Hans Oster, through imprisonment and the interrogation methods already a sick man, nevertheless maintained his pride and unrelenting attitude, which found expression in words to the interrogating SS officer at the start of the transcript: "The game is lost, the dice have fallen, yes, from the beginning I had no other intention in mind but to overthrow Adolf Hitler and his criminal regime." Wilhelm Canaris stood gray and depleted before me, certain of his fate . . . The . . . incriminating documents had still not been found. So we could coordinate our statements that Wilhelm Canaris and Hans Oster had kept me in the dark about their real coup intentions, and had pretended that the military plans to take over certain buildings in Berlin actually was an action against British and Russian commandos appearing in SS uniforms. We said good-bye with a long look among men united for many years by the same views and the same battle, forged together by the same fate, without hope for our lives . . . Once more the cleverness of Wilhelm Canaris proved successful. I was released from prison, and the interrogations and investigations began to stretch over months . . . At the last moment, as the new Gestapo arrest commando was already underway, I succeeded in fleeing, and going underground in Berlin.*

"Von Wilhelm Canaris zur NKWD," pp. 190–92.

27. Unfortunately for historians the Gestapo destroyed these documents, including Canaris's diary. On the circumstances surrounding their discovery, see Deutsch's interview with Franz Sonderegger, a Gestapo agent who was instrumental in uncovering these documents: "Contents and Fate o. Oppos. Docs. (Zossen, Canaris Diary, etc.)," in Deutsch Papers. See also Sonderegger's testimony in the Josef Müller papers, IfZ, ED 92.

28. Like the others, Dohnanyi was executed on April 9, but in his case at Sachsenhausen concentration camp.

Index